Springer Series in Advanced Manufacturing

The **Springer Series in Advanced Manufacturing** includes advanced textbooks, research monographs, edited works and conference proceedings covering all major subjects in the field of advanced manufacturing.

The following is a non-exclusive list of subjects relevant to the series:

1. Manufacturing processes and operations (material processing; assembly; test and inspection; packaging and shipping).
2. Manufacturing product and process design (product design; product data management; product development; manufacturing system planning).
3. Enterprise management (product life cycle management; production planning and control; quality management).

Emphasis will be placed on novel material of topical interest (for example, books on nanomanufacturing) as well as new treatments of more traditional areas.

As advanced manufacturing usually involves extensive use of information and communication technology (ICT), books dealing with advanced ICT tools for advanced manufacturing are also of interest to the Series.

Springer and Professor Pham welcome book ideas from authors. Potential authors who wish to submit a book proposal should contact Anthony Doyle, Executive Editor, Springer, e-mail: anthony.doyle@springer.com.

More information about this series at http://www.springer.com/series/7113

Olaf Diegel · Axel Nordin · Damien Motte

A Practical Guide to Design for Additive Manufacturing

Springer

Olaf Diegel
Design for Additive Manufacturing Lab
University of Auckland
Auckland, New Zealand

Axel Nordin
Design Sciences LTH
Lund University
Malmö, Skåne Län, Sweden

Damien Motte
Design Sciences LTH
Lund University
Lund, Skåne Län, Sweden

ISSN 1860-5168 ISSN 2196-1735 (electronic)
Springer Series in Advanced Manufacturing
ISBN 978-981-13-8280-2 ISBN 978-981-13-8281-9 (eBook)
https://doi.org/10.1007/978-981-13-8281-9

This Springer imprint is published by the registered company Springer Nature Singapore Pte Ltd.
The registered company address is: 152 Beach Road, #21-01/04 Gateway East, Singapore 189721, Singapore

Acknowledgements

A special thanks to Terry Wohlers and Prof. Ian Campbell who were instrumental in helping to create the idea for this textbook when we developed a set of DfAM guidelines for a set of courses we were giving.

A special thanks, also, to the hundreds of people in the additive manufacturing community around the world that contributed to developing much of the knowledge in this book. The more this community works together, the more we develop the knowledge to see additive manufacturing adopted by industry worldwide.

About This Book

This book attempts to impart some practical guidance on how to design parts to gain the maximum benefit from what additive manufacturing (AM) can offer. Additive manufacturing technologies are a godsend to innovation as they allow designers and inventors to quickly test out ideas to see if they work. For over thirty years, additive manufacturing has been extensively used as a rapid prototyping technology. Additive manufacturing also removes some of the traditional manufacturing constraints that have become a barrier to creativity and allows users to get real products to market without the normally high costs that can become a barrier to innovation.

When using the technologies for manufacturing, however, it should be noted that additive manufacturing does not remove all manufacturing restrictions. It, instead, replaces them with a different set of design considerations that designers must take into account if they wish to successfully use the technologies to add value to their products. Otherwise, AM can easily become a slow and uneconomical way of manufacturing products or parts.

It is also of great importance to understand that, despite much of the marketing hype over the past few decades, AM is not an "easy" technology that can make absolutely anything. It requires a good understanding of the different technologies and how to design for them. In fact, printing parts in metal, for example, can be downright hard, and the use of AM to manufacture metal parts should only be considered if the process truly adds value to the product.

In this book, you will find certain recurring themes, including the likes of support material, build orientation, post-processing, etc. These concepts are repeated because they are absolutely crucial to learning how to use DfAM in a way that makes the use of AM cost-effective and value-adding.

It is also of great importance to understand that, as these technologies are continuously evolving, new technologies, materials, and other methods for improving the processes will force us to continuously adapt our design for AM techniques. This book covers just some of the design for AM techniques that are relevant today. If you believe that any major techniques for designing, or post-processing parts, are missing, please let us know and we will try to include them in the next edition.

Much of the information presented in this book has been collected, or interpreted, from a great number of sources and, wherever possible, we have tried to acknowledge the sources of the materials and images used. If we have found sources with different, or conflicting, data, we have tended to err on the side of caution by including the overall range of values from the combined sources, reinforced by our own experiences. We have, purposefully not used an APA, Harvard, or other, reference settle in the text, as we wanted the book to be as easy as possible to read. If you believe that we have not correctly acknowledged the source of any of the data presented in this book, please let us know and we will correct in the next edition.

Some final words of advice to engineers and designers wanting to add value with additive manufacturing: Because additive manufacturing can force you to think in different ways, and allows you to realize complex shapes that could not otherwise be produced, be daring in your designs. Try new ideas just to see if they work.

People often say that, for entrepreneurship and innovation, it is important to fail fast and fail often. With additive manufacturing, we now have the ability to fail extra fast and extra often. And that's a good thing!

Olaf Diegel

Contents

About the Authors

Olaf Diegel is both an educator and a practitioner of additive manufacturing and product development with an excellent track record of developing innovative solutions to engineering problems.

In his role as professor of product development, in the department of design sciences of the faculty of engineering at Lund University, in Sweden, he is heavily involved in all aspects of product development and is widely published in the areas of additive manufacturing and rapid product development. In his consulting practice, he develops a wide range of products for companies around the world. Over the past three decades, he has developed over 100 commercialized new products including innovative new theater lighting products, security and marine products, and several home health monitoring products and, for this work, has received numerous product development awards.

Over the last 25 years, Olaf has become a passionate follower of 3D printing (additive manufacturing). In 2012, Olaf started manufacturing a range of 3D-printed guitars and basses that has developed into a successful little side-business (and gives Olaf the therapy he needs in allowing him to make things that are a blend of high-technology and traditional handcrafting).

Axel Nordin holds an M.Sc. in mechanical engineering from Lund University, Sweden, and a Ph.D. of engineering from the division of machine design. He has participated in several government-funded research projects. His work is mainly concerned with studying aspects of integrating complex morphologies into bespoke products, such as computational, manufacturing, structural, and usability challenges, as well as in topology optimization and generative design, and how these can be applied in the field of design for additive manufacturing.

Damien Motte is an associate professor at the division of product development with the School of Engineering at Lund University, Sweden. He received a Ph.D. from the same division, a research master from the Industrial Engineering Laboratory at École Centrale Paris, France, and an M.Sc. in Industrial Engineering at École des Mines d'Albi, France. His area of research focus includes alternative engineering design, design for additive manufacturing, and product development methodologies.

Acronyms

The following are acronyms used in the field of additive manufacturing and 3D printing:

3DP	3D printing
3MF	Additive manufacturing file format used to describe color, textures, materials, and other characteristics of a 3D model. Ongoing development of the file format is led by the 3MF Consortium, which was initiated by Microsoft and other companies in 2015
ABS	Acrylonitrile butadiene styrene; a thermoplastic polymer with high impact resistance and toughness
AJP	Aerosol jet printing
AM	Additive manufacturing
AMF	Additive manufacturing file format for communicating additive manufacturing model data including a description of the 3D surface geometry with native support for color, materials, lattices, textures, constellations, and metadata
B2B	Business to business
B2C	Business to consumer
BAAM	Big area additive manufacturing. A large-scale form of material extrusion
CAD	Computer-aided design; the use of computers for the design of real or virtual objects
CAE	Computer-aided engineering; CAE software offers capabilities for engineering simulation and analysis, such as determining a part's strength or its capacity to transfer heat
CAM	Computer-aided manufacturing; typically refers to systems that use surface data to drive CNC machines, such as digitally driven mills and lathes, to produce parts, molds, and dies

CLIP Continuous liquid interface production
CNC Computer numerical control; machines equipped with CNC capabilities include mills, lathes, and flame cutters
CT Computed tomography; CT scanning is a method of capturing the internal and external structures of an object using ionizing radiation. A CT scan creates a series of two-dimensional grayscale images that can be used to construct a 3D model
DDM Direct digital manufacturing. See definition in Glossary
DED Directed energy deposition. See definition in Glossary
DfAM Design for additive manufacturing
DLP Digital light processing, a technology developed by Texas Instruments
DMD Direct metal deposition
DMLS Direct metal laser sintering. See definition in Glossary
DMP Direct metal printing
EBAM Electron beam additive manufacturing
EBM Electron beam melting
EDM Electrical discharge machining; a method of machining that removes material with a series of electrical current discharges between a tool electrode and a workpiece
FDM Fused deposition modeling
FFF Freeform fabrication; another name for additive manufacturing
GF Glass filled
HIP Hot isostatic pressing
ISO International Standards Organization; more widely known as the International Organization for Standardization
LBM Laser beam melting
LENS Laser engineered net shaping
LOM Laminated objet manufacturing
LS Laser sintering
MCAD Mechanical computer-aided design; the use of CAD to design mechanical parts and assemblies
MEMS Microelectromechanical systems
MJF Multi-jet fusion. A powder bed fusion processed developed by HP
MRI Magnetic resonance imaging; alternative to CT scanning that offers better soft-tissue contrast; MRI does not use ionizing radiation
NSF National Science Foundation; a US government funding agency
OEM Original equipment manufacturer
PA Polyamide; a family of thermoplastic polymers often used for powder bed fusion systems
PAEK Polyaryletherketone; a high-melting-temperature thermoplastic polymer; a member of the polyaryletherketone family
PBF Powder bed fusion

PBT	Polybutylene terephthalate; a strong thermoplastic polymer used as an insulator and is resistant to solvents
PC	Polycarbonate; a family of thermoplastic polymers that are highly formable with high impact resistance
PCL	Polycaprolactone; biodegradable polyester used to produce specialty polyurethanes
PEEK	Polyetheretherketone; a high-melting-temperature thermoplastic polymer; a member of the polyaryletherketone family
PEI	Polyethylenimine; a polymer used for adhesives, detergents, and cosmetics
PEKK	Polyetherketoneketone; a high-melting-temperature thermoplastic polymer; a member of the polyaryletherketone family
PHA	Polyhydroxyalkanoate; polyesters produced naturally from bacterial fermentation of lipids or sugar; biodegradable and used to produce bioplastics
PIM	Plastic injection molding; popular method of molding parts from thermoplastic materials such as polypropylene, polyamide (nylon), polycarbonate, ABS, polyethylene, and polystyrene
PLA	Polylactic acid; a thermoplastic polymer that is biodegradable and often derived from renewable sources
PLLA	Poly-L-lactic acid (also referred to as polylactic acid); biodegradable polyester made from corn starch, sugarcane, or tapioca roots
PMMA	Polymethyl methacrylate; a thermoplastic polymer that is used in Voxeljet's binder jetting process
PP	Polypropylene; a thermoplastic polymer used in a wide variety of manufacturing applications
PPS	Polyphenylene sulfide; an organic polymer often used for making filter fabric
RE	Reverse engineering. See definition in Glossary
RM	Rapid manufacturing
RP	Rapid prototyping. See definition in Glossary
SFF	Solid freeform fabrication; another name for additive manufacturing
SHS	Selective heat sintering
SL	Stereolithography
SLA	Stereolithography or stereolithography apparatus
SLM	Selective laser melting
SLS	Selective laser sintering
SMEs	Small- and medium-sized enterprises
STEAM	Science, technology, engineering, art, and mathematics
STEM	Science, technology, engineering, and mathematics; often used in association with education policy and curriculum development in schools to help improve competitiveness

STL File format for 3D model data used by machines to build physical parts. STL is the de facto standard interface for additive manufacturing systems. STL originated from the term stereolithography. The STL format uses triangular facets to approximate the shape of an object, listing the vertices, ordered by the right-hand rule, and unit normals of the triangles, and excludes CAD model attributes

TPU Thermoplastic polyurethane; a class of polyurethane plastics (thermoplastic elastomers) that share properties of elasticity, transparency, and resistance to oil and grease

WAAM Wire arc additive manufacturing

Chapter 1
Introduction to Additive Manufacturing

1.1 What Is Additive Manufacturing?

Additive manufacturing (AM) encompasses a range of technologies that allows physical components to be made, from virtual 3D models by building the component layer-upon-layer until the part is complete.

In comparison with subtractive manufacturing processes, in which one starts with a block of material and removes any unwanted material (either by carving it by hand, or by using a machine such as a mill, lathe or CNC machine) until one is left with the desired part, additive manufacturing starts with nothing and builds the part one layer at a time by 'printing' each new layer on top of the previous one, until the part is complete (Fig. 1.1). Depending on the particular technology used, the layer thickness ranges from a few microns up to around 0.25 mm per layer, and a range of materials are now available for the different technologies. These are discussed in the next chapter.

The very earliest concepts related to additive manufacturing date back to the end of the 19th century, and early 20th century, with the introduction of layer-based topographical maps as 3D representations of terrain, together with a number of methods for using these topological models to produce 3D maps by, for example, wrapping a paper map over the topological models to produce a 3D model of the terrain.

Photosculpture, which also originated towards the end of the 19th century, and which used a series of different photographs taken from different angles around the object that were then used to carve out the object using each different angled picture as a template, so an initially subtractive process, also had several proposed methods for creating the models using photosensitive materials.

Modern additive manufacturing saw its origins in the mid-20th century with a patent, in 1951, by Otto John Munz which could be considered the origin of the modern stereolithography technique. It consisted, essentially, of a series of layered 2D transparent photographs printed on photosensitive emulsion stacked on top of each other. He developed a system for selectively exposing the transparent

O. Diegel et al., *A Practical Guide to Design for Additive Manufacturing*, Springer Series in Advanced Manufacturing,
https://doi.org/10.1007/978-981-13-8281-9_1

Fig. 1.1 Subtractive manufacturing, where one builds a part by removing material from a block, versus additive manufacturing, in which one builds the part layer-upon-layer

photo-emulsion in a layer-wise fashion in which each layer was exposed with a cross section of an object. Much like a modern stereolithography machine, the build platform on which the part was being built was gradually lowered, and the next layer of photo emulsion and fixing agent was created on top of the previous layer. Once the printing process was finished, the result was a solid transparent cylinder containing a 3D image of the object. A weakness of this system was that the final real three-dimensional object had to be manually carved or photo-chemically etched out of the cylinder as a secondary operation.

The following decades saw the development of a succession of new techniques including those of Swainson who, in 1968, proposed a process to directly fabricate a plastic pattern through the selective 3D polymerization of a photosensitive polymer at the intersection of two laser beams (with the patent assigned to the Formigraphic Engine Corporation). Work was also undertaken at Battelle Laboratories, called Photochemical Machining, in which an object was formed by either photochemically crosslinking or degrading a polymer through the simultaneous exposure to intersecting laser beams.

In 1971, Ciraud proposed a powder process that can be considered the father of modern direct deposition AM techniques such as powder bed fusion, and in 1979, Housholder developed the earliest equivalent of a powder-based selective laser sintering process. In his patent, he discussed sequentially depositing planar layers of powder and selectively solidifying portions of each layer. The solidification could be achieved by using heat and either a selected mask, or by using a controlled heat scanning process such as a laser.

Other notable early additive manufacturing developments include those of Hideo Kodama, of the Nagoya Municipal Industrial Research Institute in Japan, who developed a number of stereolithography related techniques, and the work of Herbert who,

Fig. 1.2 Examples of early AM parts by Householder, Kodama, Herbert and Manriquez Frayre and Bourell [1]. Courtesy of Ismail Fidan, Dave Bourell, and IS&T: The Society for Imaging Science and Technology [2]

in parallel with Kodama, developed a system that directed a UV laser beam onto a photopolymer layer, by means of a mirror system on an x–y plotter, to scan a layer of the model. The build platform and layer were then lowered by 1 mm into a vat of resin and the process was repeated. Examples of Householder's, Kodama's and Herbert's parts can be seen in Fig. 1.2.

Commercial additive manufacturing, as we know it today, with the development of commercially available system did not really begin until 1986, with Charles W. Hull's stereolithography patent. The patent was originally owned by UVP Inc. and the company licensed the technology to their former employee, Charles Hull, who used it to found the start-up 3D Systems. This development saw the first commercial SLA machine appearing in 1988 and, since then, almost every year has seen an exponential rise in available systems, technologies and materials.

Even the terminology relating to additive manufacturing has changed a lot over the last three decades. For most of the 1990s, the principal term used to describe the layer-upon-layer manufacturing technologies was rapid prototyping (RP), because the principal use of the various available technologies was to make concept models and pre-production prototypes. Some other terms that have also been used over the years include Solid Freeform Fabrication (SFF) and Layer Manufacturing.

In early 2009, however, the ASTM International Committee F42 on Additive Manufacturing Technologies tried to standardize the terminology used by the industry and, after a meeting in which many industry experts debated the best terminology to use arrived at the term 'Additive Manufacturing' which, today, is considered the standard terminology used by industry.

In their ASTM F2792 10e1 Standard Terminology for Additive Manufacturing Technologies document they defined additive manufacturing as:

> the process of joining materials to make objects from 3D model data, usually layer upon layer, as opposed to subtractive manufacturing methodologies, such as traditional machining.

Unlike subtractive manufacturing, where material is removed from a larger block of material until the final product is achieved, most additive manufacturing processes do not yield excessive waste material. If a part is properly 'designed for AM', and one is comparing it to a single part produced through conventional manufacturing, it also typically may not require the large amounts of time needed to remove unwanted material thus, potentially, reducing time and costs, and producing relatively little waste. This, however, should not be misconstrued as AM being able to always make cheaper parts than conventional manufacturing. In many cases it is, in fact, the opposite, because AM is a relatively slow and expensive technology. This is discussed in greater detail in the chapter on the economics of AM. But this depends very much on the AM technology in use and the many possible design parameters that can be used and is, in part, what this book is about.

It should be noted, however, that while industry has, generally, adopted the additive manufacturing term, much of the popular press and media continues to refer to additive manufacturing as 3D printing, as this is a term more easily understood by the general public. Some consider the term 3D printing to be focussed on lower-cost hobby desktop 3D printers, and additive manufacturing focussed on higher-end industrial production systems. In this book, when talking about the process, we use the two terms interchangeably.

1.2 The Additive Manufacturing Process Chain

All additive manufacturing begins with the creation of a virtual 3D CAD model. Almost any 3D computer aided design (CAD) software can be used. This CAD model must, however, be in the form of a fully enclosed "watertight" volume (which means that a model of a cube, for example, must include all 6 faces, and have no gaps at the seams). If one of the faces of the model is missing, or there are gaps, it represents an infinitely thin surface which cannot be printed (though, depending on how bad the errors are, some AM software can be used to automatically fix the model) (Fig. 1.3).

The CAD file is then converted into a file format that can be understood by the AM machine. Typically, today, the most commonly used file format is an STL file (standard triangle language, stereolithography, or standard tessellation language) which is a format that converts the original CAD file into a triangulated file. The higher the resolution of the STL file, the more triangles it contains, so the better the quality of the model (Fig. 1.4).

Some new additive manufacturing file formats, including AMF (Additive Manufacturing File Format) and 3MF (3D Manufacturing Format), have recently been proposed which vastly improve the somewhat antiquated STL format, as they add more information to the file, including colour and material, and allows the use of curved triangles to improve model quality. At the time of publication, 3MF appears to be gaining considerably more traction than the AMF format.

3MF, or 3D Manufacturing Format, is a file format developed and published by the 3MF Consortium. It is an XML-based data format designed specifically for AM, and it includes information about materials, colours, and other information that cannot be represented in the STL format. The 3MF file format is in the process of being adopted by companies including Autodesk, Dassault Systems, Netfabb, Microsoft, SLM Solutions, HP, Shapeways, Materialise, 3D Systems, Siemens PLM Software and Stratasys.

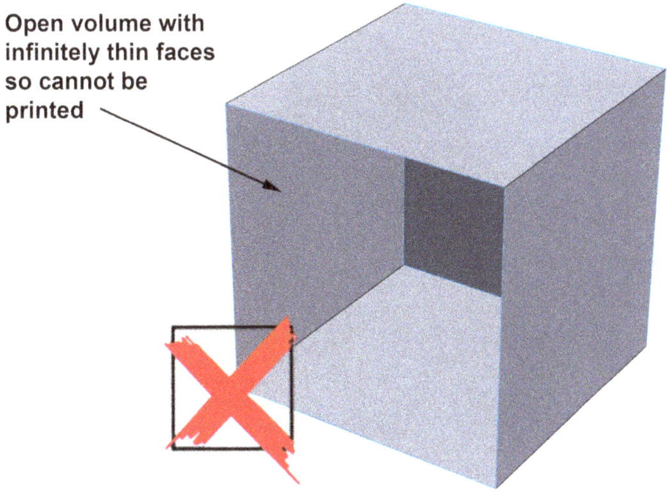

Fig. 1.3 A non-watertight open surface model that cannot be printed because it is made up of an open set of infinitely thin faces

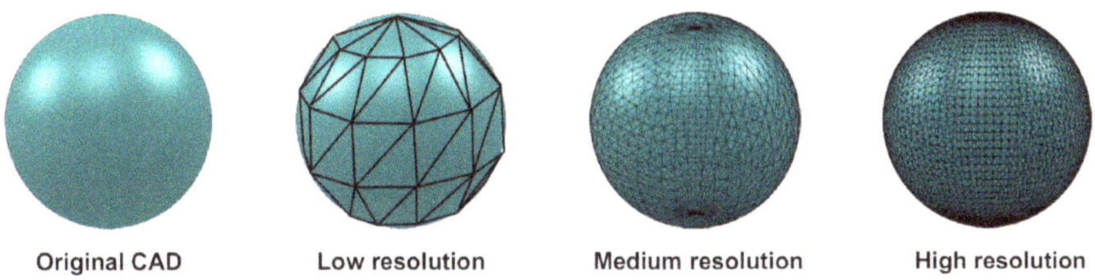

Fig. 1.4 Examples of STL file resolution

Some researchers are working on printing directly from the native CAD formats, which is the most potentially promising option as it avoids all file conversion which, by its nature, will tend to degrade the quality of the model. Today, however, the vast majority of AM systems and desktop 3D printers, today, still use STL files.

The STL file produced by the CAD software is then opened in the AM machine's software, and the model is placed on the software's virtual build platform (the platform on which the part will be printed) in the most suitable orientation for printing. The print orientation can affect both the surface quality, and the strength of the final part. Some processes, for example, produce highly anisotropic parts in which there is a weakness between the layers of the part, or vertical direction of the print. Other processes use support material to allow overhanging parts to be printed. These aspects can be critical to additive manufacturing, so will be discussed in greater detail in following chapters (Fig. 1.5).

The AM machine's software then slices the STL files into thin layers, and some software also let you set the other print parameters including print resolution (layer thickness), material, fill patterns, speed, etc.

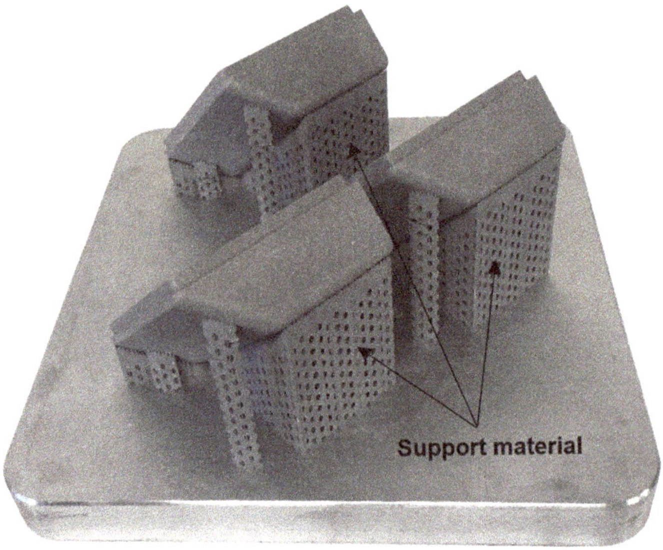

Fig. 1.5 Example of the support material required by some processes

Once the software has sent the part build instructions to the machine, it starts to build the part layer upon layer. How it builds each layer, and what material it uses, depends on the particular technology being used, and most of these common additive manufacturing technologies will be discussed in the next chapter (Fig. 1.6).

After the machine has finished printing the part, they are removed from the machine and post-processed. Post-processing almost always includes cleaning the part of left-over powder or resin, removing support material and, in many cases, might include further processing such as machining, if a surface requires a finer finish than the AM machine can provide, infiltration to make the part stronger, heat treatment for metal parts, or colouring and painting if the part needs to be in a colour other than that provided by the AM material.

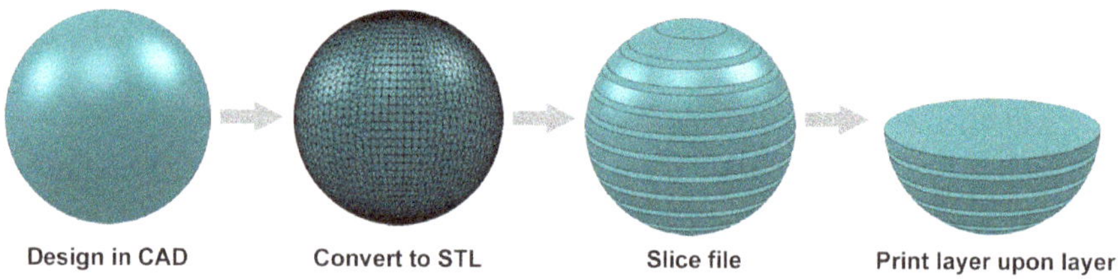

Fig. 1.6 The additive manufacturing process chain

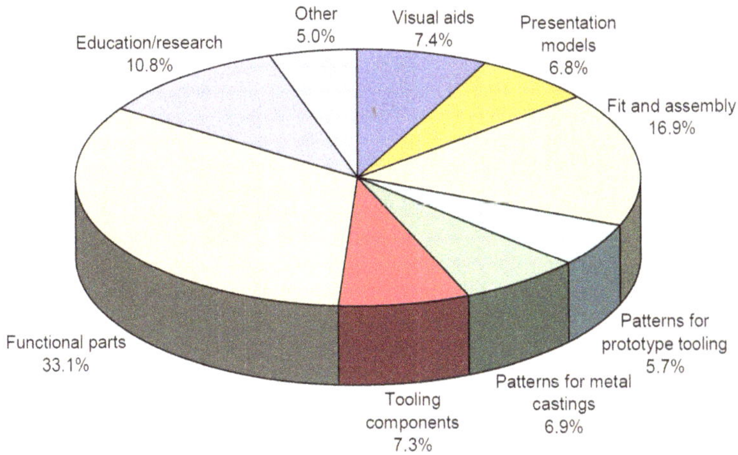

Fig. 1.7 Application areas that AM is currently being used for, courtesy of Wohlers Associates

1.3 Current Usage of Additive Manufacturing

Over the past 30 years, AM has grown to be used in an ever increasing number of application areas. The Wohlers Report, a leading annual 'state of the industry' report, undertakes a yearly survey to find out what AM is being used for. The 2018 Wohlers Report gives the following data on areas of use for AM as shown in Fig. 1.7.

It is interesting to note that, although 43.9% of applications are in the rapid prototyping realm (including fit and assembly, functional models, presentation models and visual aids) the use of AM to produce real parts, both directly or indirectly, now represents over 56% of usage. This includes patterns for prototype tooling, patterns for metal casting, tooling components, and direct part production. Wohlers expects this percentage to grow substantially over the next few years as more and more industries adopt AM as part of their growing manufacturing arsenal.

Since its inception, AM has grown on two distinct fronts. The last decade, in particular, has seen good improvements on high-end machines capable of producing excellent quality parts in a number of materials. At the same time, there has been enormous growth in the DIY and desktop 3D printer community with a wide range of entry level additive manufacturing systems with prices ranging from a few hundred to a few thousand dollars. Entire communities (the reprap, fablab, and makerbot communities, for example) have developed, often with an open-source approach to sharing knowledge on additive manufacturing which has greatly benefitted the industry.

It is also interesting to look at the growing spread of industries that are using additive manufacturing (Fig. 1.8).

It is only over the last few years that additive manufacturing has improved in quality to the extent that some companies have started to use it as a viable production

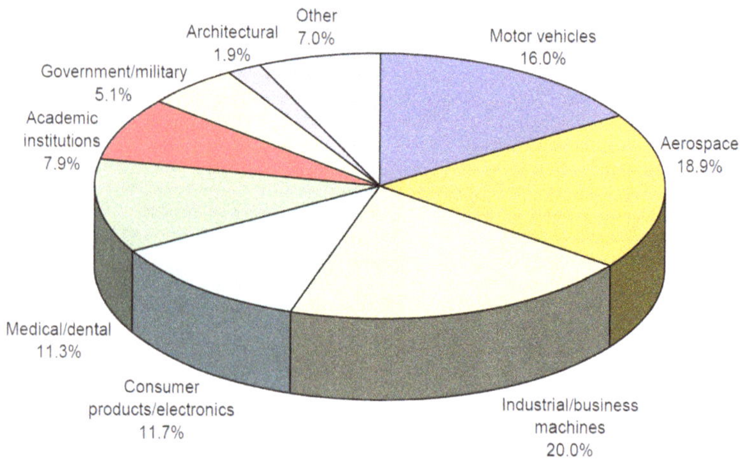

Fig. 1.8 Industries using additive manufacturing, courtesy of Wohlers Associates

technology. As new polymer and metal materials are developed and the speed and precision of the machines further increase, more additive manufacturing machines are likely to find their way into mainstream production lines.

Additive manufacturing also has a number of qualities that give it the ability to manufacture parts that cannot be made by traditional manufacturing techniques. Understanding this is vital to understanding when, and when not, to use AM. It is also important to note that AM will never completely replace traditional manufacturing. It is a complimentary technology that, if used because of the value it can add, and if the parts it produces are specifically designed for AM, then it can add great value to the company. Some of the advantages AM has over conventional manufacturing are listed below.

1.4 The Advantages of Additive Manufacturing

AM can be an expensive process, so in order for its use to be profitable as a production method, it must bring added value to a product. This can either be through reducing life cycle costs for the product or through enabling a higher price to be charged to the customer. This is achievable in a number of ways as described below.

1.4.1 Part Complexity

Additive manufacturing enables the creation of parts and products with complex features, which could not easily have been produced via subtractive or other traditional manufacturing processes. With conventional manufacturing, the more geometrically complex a part becomes the more expensive it becomes to manufacture and, at a certain point, it becomes impossible to manufacture. AM works in the opposite way: The more geometrically complex the part is, the more suitable it is for AM, as it costs no more to produce a complex part than a simple one (there are some exceptions to this, particularly if support material is hard to remove, and this is discussed later in this book). If a part is very simple, however, AM can become an expensive way of producing the part compared to traditional manufacturing.

With AM, more complex shapes can be created, both for external forms and internal structures. This can result in improved product performance and/or increased aesthetic attraction. The former can be translated into lower running costs and the latter into increased price (Fig. 1.9).

With traditional injection molded or die-cast parts, for example, the parts must be removable from the die in which they are made and must therefore be designed in such a way that this can be done. For simple 'open and close' die parts this is not a problem but, as the complexity of the parts increases, a number of 'moving cores' becomes necessary which can greatly increase the complexity, and cost, of the tool. And, past a certain level of complexity, the parts cannot be manufactured at all, or must be broken down into a number of smaller components that then need to be assembled to make the final component possible. With AM, this is no longer a problem as the complex part can be manufactured directly (Fig. 1.10).

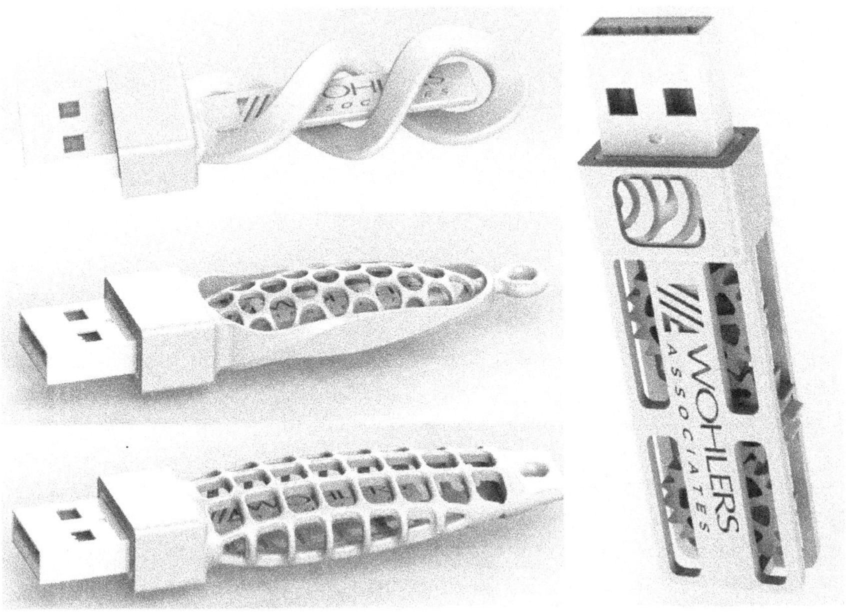

Fig. 1.9 Examples of geometrically complex forms that would be hard, or impossible to make with traditional manufacturing

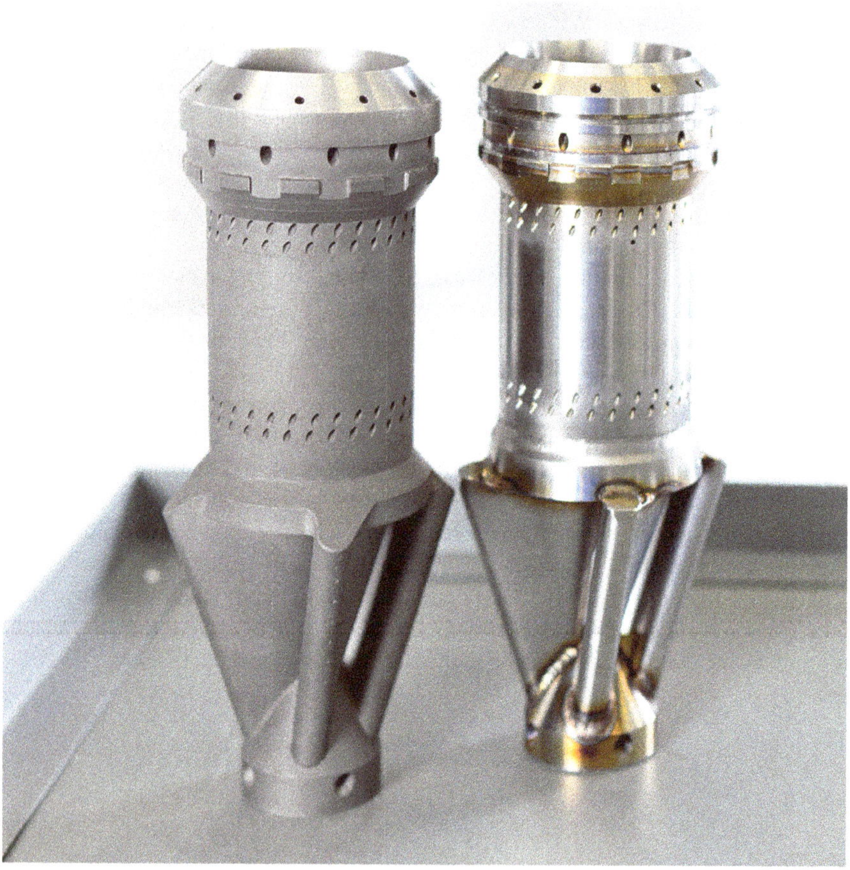

Fig. 1.10 Siemens turbine burner, courtesy of Siemens. This part would be impossible to machine, or cast, as a single piece, but can be made through additive manufacturing

The Siemens turbine burner metal part shown above, for example, could not easily be machined, or cast, because there is no way of removing the internal part of the die from the component or of machining the interior surfaces. The only way to manufacture the component with conventional manufacturing is to break it down into several sub-components that then get welded together. There are, however, great advantages if it is possible to manufacture it as a single component. Additive manufacturing does not suffer from these particular complexity restrictions. The complexity of the part does not, generally, affect whether it can be made, or even greatly affect its cost.

The Siemens burner went from 13 machined parts joined together by 18 welds, down to a single AM part. Because of the improved cooling that AM allowed, it was also possible to remove the need for a thermal barrier coating that was required on the conventional design. The pilot gas feed was integrated into the structure this removing the need for an external pilot gas feed pipe. The part weight also went from 4.5 down to 3.5 kg for the AM produced part.

It should be noted that additive manufacturing does not remove all manufacturing restrictions. It, instead, replaces them with a different set of design considerations that designers must take into account if they wish to successfully use the technologies. These design guidelines and constraints are what this book is about.

A simple example of the type of restrictions that additive manufacturing suffers from is the inability to manufacture entirely enclosed hollow volumes. A completely sealed hollow sphere, for example, is still impossible for current additive manufacturing technologies to make because there is no way of removing the excess resin, powder, or support material, from the inside of the sphere. One must therefore leave a hole of a minimum diameter in the part in order to remove the excess material from inside the part.

These new design considerations are, however, much less restrictive than traditional manufacturing technologies, and easier for designers to both understand and comply with without them affecting design intent in a major way.

1.4.2 Instant Assemblies

With additive manufacturing it is possible to manufacture complex interlocked moving parts in ready-made working assemblies. Though two components may be permanently linked together, they are made as a single component and come out of the machine assembled and ready to work.

Figure 1.11, for example, shows a foldable guitar stand that was printed using a powder bed fusion technology. The entire guitar stand is, however, manufactured in a single operation with no assembly whatsoever required. If the guitar stand were to be manufactured using traditional manufacturing methods, it would require, at least, sixteen components and an assembly procedure to attach all the separate components together.

To additively manufacture complete assemblies, the designer needs to leave a minimum gap between the moving components. The material in these gaps is not processed by the AM machine so, after the part is finished, can be removed to leave the surrounding components free to move. The size of this gap varies from process to process but is, generally, in the order of a fraction of a millimetre. It is important to understand that the required gap between moving parts is considerable, by engineering standards, so if moving parts with tight engineering fits are required, it is unlikely that AM would be able to print them in an assembled configuration. This is discussed in greater detail in the general guidelines for part consolidation section of this book.

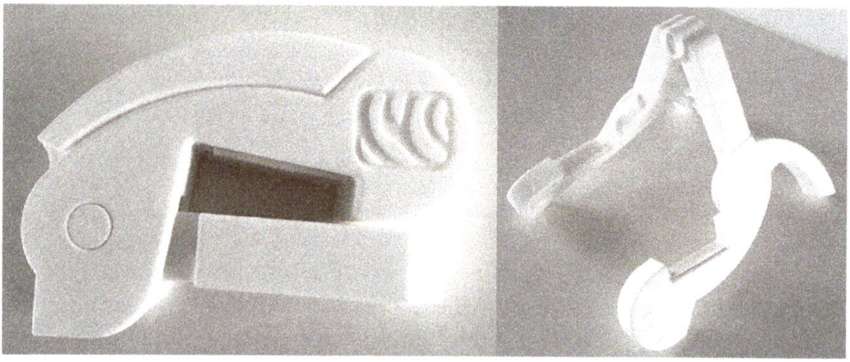

Fig. 1.11 Foldable guitar stand manufactured as a single component using polymer powder bed fusion

1.4.3 Part Consolidation

Part consolidation is when several simpler parts are replaced by a single more complex AM part. Figure 1.12 shows an example of part consolidation. This reduces assembly and inventory costs. A later chapter of this book focuses on the guidelines for AM part consolidation.

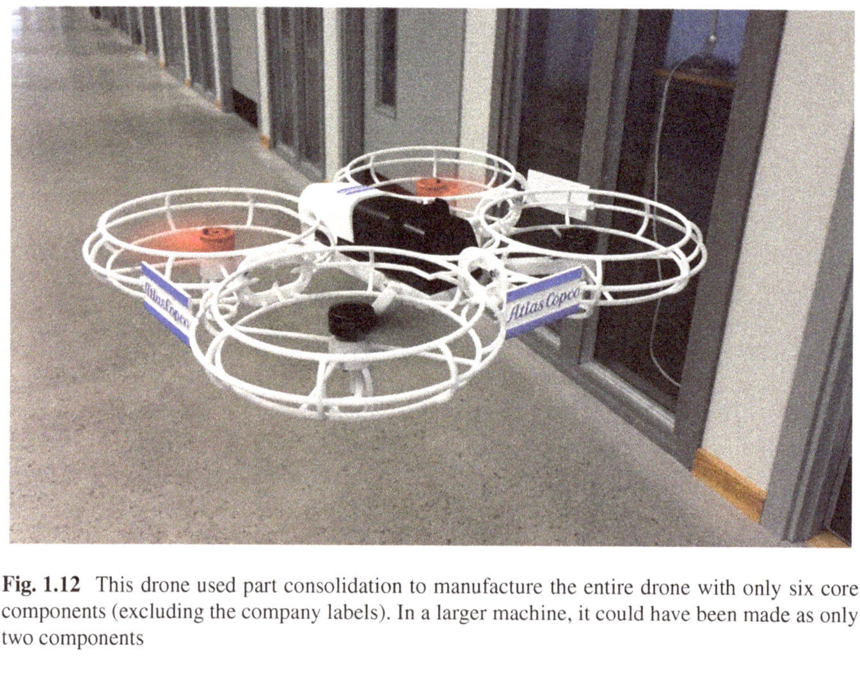

Fig. 1.12 This drone used part consolidation to manufacture the entire drone with only six core components (excluding the company labels). In a larger machine, it could have been made as only two components

1.4.4 Mass Customization

With additive manufacturing parts can be made on demand, as there is no longer a long lead-time to get tooling produced. The tooling required for conventional mass-manufacturing typically takes from a few weeks to several months to produce. This feature of additive manufacturing has a great impact on new product time-to-market, and on the ability to easily produce model changes throughout the life of a product. It also has implications in stock control: As components can be made on the spot, companies may no longer need to hold large stock of spare parts as they simply manufacture the parts when needed. This feature of additive manufacturing is often referred to as 'manufacturing-on-demand'.

From a product design perspective, it also means that every component made can be completely different to the others in a production run without significantly affecting the manufacturing cost or manufacturing time. This opens the door to mass-customization in which, though mass-manufactured, each product can be customized to each individual customer. This is already beginning to happen in industries including hearing aids, dental crowns, implants, medical prostheses (Fig. 1.13), customized orthotic shoe inserts, and the high-end interior design and fashion industries.

For this new way of designing products to be used effectively, the product design and the computer aided design industries will need to develop new methods for integrating personalized customer data into their designs. This development has already started, particularly in the hearing aid and the dental industries, in which specialized software exists to automate the processes of patient data acquisition. The patient's personalized data is acquired, usually through a laser scanning process, and the software then automates the process of cleaning of the data, shelling the part, inserting mounting points for electronic components, etc. This increased automation in CAD software now needs to be extended to encompass other industries, including consumer product industries.

1.4.5 Freedom of Design

One of the greatest advantages additive manufacturing gives the designer over traditional manufacturing is freedom of design. Because of constraints of traditional manufacturing technologies, a product, which the designer may have originally envisioned as having a certain aesthetic and functionality, may need to be compromised so that it can be cost-effectively made. Most designers are quite accustomed to hearing the response of "it cannot be made like that" from manufacturing engineers. They may then need to compromise their design to the extent that the product loses the essence that truly embodies the designers vision. With additive manufacturing, complexity and geometry often no longer affect manufacturability. Almost anything the designer imagines can be made precisely as the designer conceived it (with the

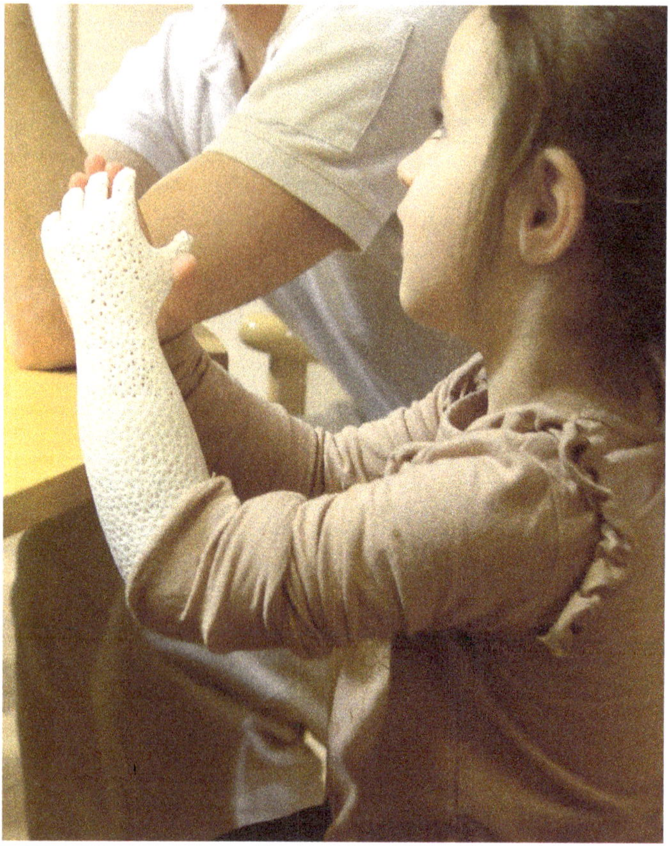

Fig. 1.13 Individually customized prosthetic

proviso that the parts may require substantially more labour in the post-processing stage).

Though this is directly related to product complexity, discussed above, we are, here, talking about the ability it gives designers and engineers to be less restricted by manufacturing constraints, and thus allows them to innovate in a way that may not have been previously possible (Fig. 1.14).

1.4.6 Light-Weighting

Topology optimisation is a method of removing as much material as possible from a part, while maintaining sufficient mechanical properties. It consists in performing a finite element analysis (FEA) and then iteratively removing unnecessary material.

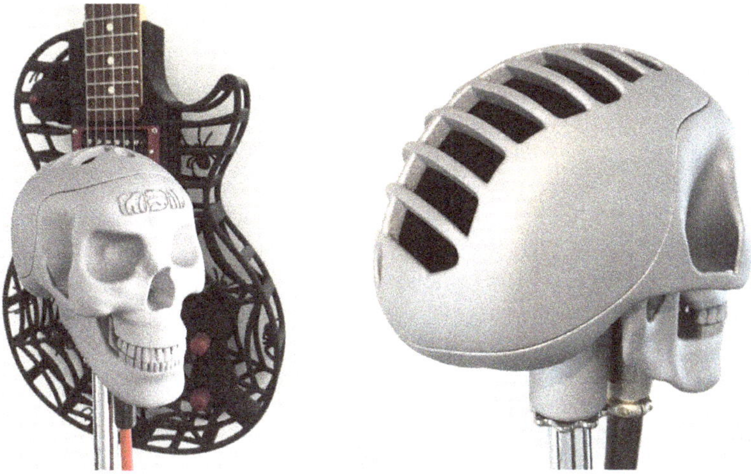

Fig. 1.14 Geometrically complex skull-shaped microphone. This products could not have been cost-effectively made using traditional manufacturing methods

Topology optimisation has been available for a long time, but complex designs could not be manufactured with conventional manufacturing. AM, however, is capable of making most complex designs from topology optimisation. This opens up a whole new engineering area dedicated to making lighter products. In the context of aeronautics, for example, any weight savings can represent large savings in fuel.

In the example of the aircraft seat, in Fig. 1.15, it would require a block weighing 16.2 kg to be machined down into the finished product, weighing 4.1 kg. In contrast, the topology optimised version weighs only 3.1 kg.

A later chapter of this book focuses on using topology optimisation in the most effective way.

Fig. 1.15 CNC-machined aircraft seat frame (left), rough topology optimized version (centre), and finished design (right)

1.4.7 On-Demand Manufacturing

The term "on-demand manufacturing" or "manufacturing on demand" is a manufacturing process in which goods are produced as and when required and, in the context of additive manufacturing, where required. With conventional manufacturing, an assembly line produces large quantities of products, which are then kept in stock until they are ready for shipping to their intended location. With on-demand manufacturing, products are made only once the customer's order comes in, and only as many products as are immediately required are made. If AM is being used, the AM system can be located near where the customer is, and the products can be made locally, rather than shipped around the world.

One can see the potential this has in entirely disrupting our existing supply chains. The benefits of this new supply chain are increasingly evident. It has the potential to lead to:

- Cost savings by eliminating or significantly reducing inventory requirements. Warehouses full of spare parts and stock could become a thing of the past. Instead of a physical inventory of parts, we now switch to keeping a digital inventory of parts, and only make the parts physical the moment they are required.
- Digital files also provide the ability to quickly produce new product iterations at little to no additional cost. And as keeping stock of old product is no longer necessary, these changes can be instantly implemented.
- With a single source for a variety of parts, businesses that use 3D printing contract manufacturers deal with less risk, more control and added agility in relation to their product lifecycle.
- Local facilities can 3D print designs on-demand from files sent across the globe, or they can print securely from a nearby supplier.

Prior to the industrial revolution, the supply chain was extremely small. Most production was local to where the products where being sold, and transporting products over long distances was often not feasible. The supply chain often began with the end-user making the product for themselves, perhaps with some bartering for parts from local trades-people and, if the product was good, it could then be sold to the rest of the market which, in most cases, was local to the village the maker lived in (Fig. 1.16).

Fig. 1.16 The supply chain before the industrial revolution

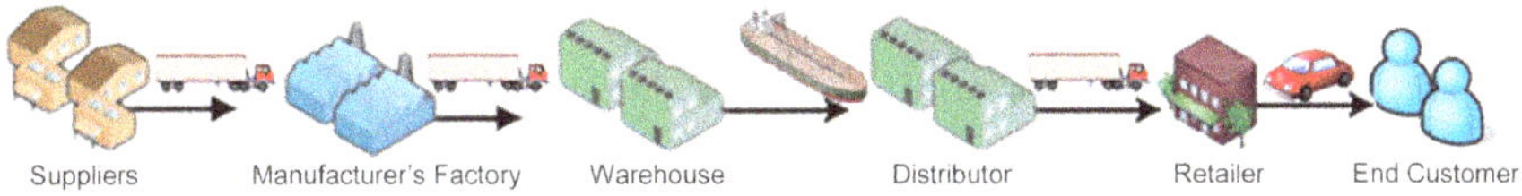

Fig. 1.17 The supply chain today

Fig. 1.18 The supply chain of tomorrow

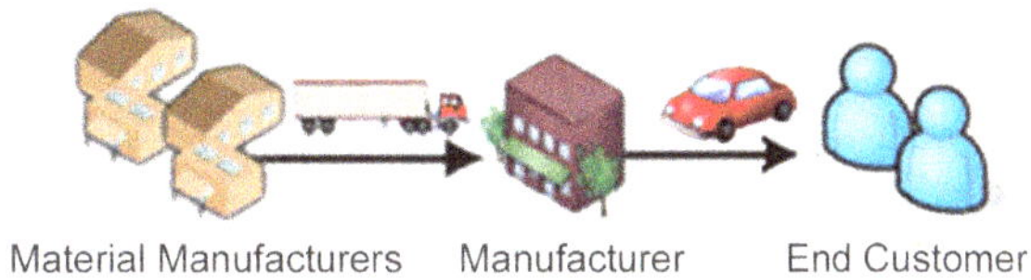

In contrast, once mass-production began during the industrial revolution, products began to be mass-manufactured in a location, and then transported to the end-user somewhere in the world, through a complex supply chain of middle-men. Each of these middle-men adds a percentage cost to the product to make their margins, and the environmental foot print caused by all the transport and storage of the products can now be seen to be having an impact on the world we live in (Fig. 1.17).

Perhaps, as the industrial adoption of AM grows, we may return to a simpler and more compact supply chain model that more closely resembles the pre-industrial revolution supply chain (Fig. 1.18).

Chapter 2
Additive Manufacturing Technologies

This chapter describes the main additive manufacturing technologies in use today. Note, however, that new technologies are continuously appearing on the market, so this space is in a continuous state of flux. Some of the technologies also have minor variants such as, for example, a material extrusion which might have one, two, or even more, extrusion nozzles. The intent of this chapter is not to cover every single technology in existence but, rather, to give the novice an overall understanding how each category of AM technology works.

The terminology used below tries to conform to the established ASTM F42 terminology for additive manufacturing technologies which, broadly, divides them based on how the material is solidified into the desired component.

2.1 Material Extrusion

Material extrusion based additive manufacturing technologies, the most common of which is known as Fused Deposition Modelling (FDM), work by extruding a fine filament of material, usually polymer, and tracing out each slice of the model with that material. The build platform then moves down a fraction of a millimetre, and the next slice is then deposited on top of the first and, as the plastic is in a molten state, is bonded to the previous layer.

The easiest way of understanding the process is to think of it as a computer controlled hot-glue gun that is used to draw each slice of the model. Material extrusion systems often have a second nozzle that deposits a different sacrificial 'support' material for each layer of the model. The support material is there to produce a surface for any overhanging bits to be printed onto. After the model is complete, the support material is then removed by either breaking it off or dissolving it (Fig. 2.1).

There are now a number of material extrusion systems on the market capable of printing with composite materials, including continuous strand carbon fibre, chopped strand composites (carbon fibre, Kevlar, glass gilled, etc.). These either use a filament

O. Diegel et al., *A Practical Guide to Design for Additive Manufacturing*, Springer Series in Advanced Manufacturing,
https://doi.org/10.1007/978-981-13-8281-9_2

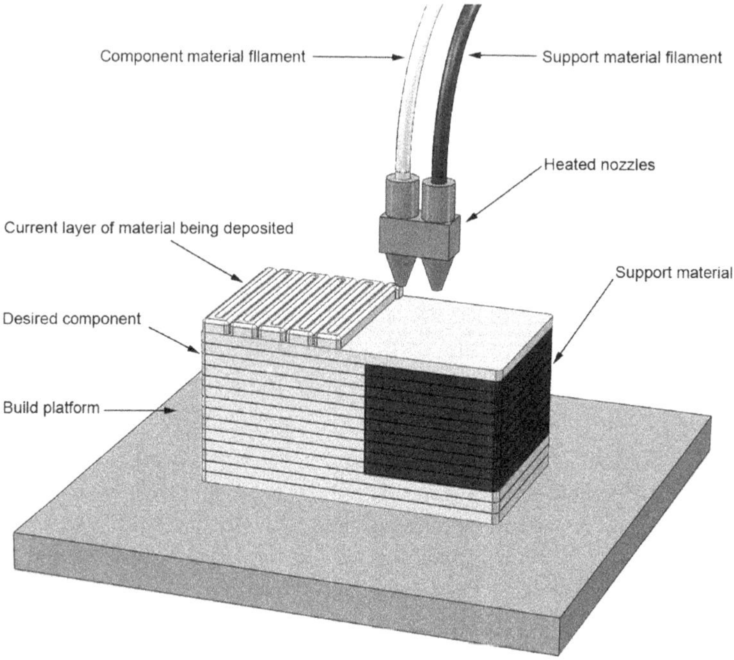

Fig. 2.1 Material extrusion based additive manufacturing system

in which the polymer is mixed with chopped strands of the composite filler material, or co-extrude a polymer with a continuous strand of fibre.

Metal material extrusion systems are now becoming available, in which a filament, or rod, consisting of polymer filled with metal powder (often with around 80% metal powder content) is extruded to form composite metal/polymer parts that can then be sintered in a furnace to burn off the polymer and fuse the metallic powder together to produce metal parts. Parts can shrink by up to 20% on sintering, so special consideration needs to be given to part design, in order to make this shrinkage uniformly controllable (Fig. 2.2).

One of the challenges with material extrusion systems is that the parts it produces tend to be anisotropic (more so than other AM processes), in such a way that the parts are weaker in the vertical 'Z' direction than in the horizontal direction (X and Y). This is because the bond between the layers of material is slightly weaker than the plastic that forms the layers. It can be thought of as a wood grain, where the wood is strong when force is applied perpendicular to the grain, but weak when a force is applied parallel to the wood grain. So, in general, for parts that will be used largely under compression, material extrusion is a suitable rapid manufacturing process, but less so for parts that may be under tension, which might cause the layers to delaminate. It should also be noted that material extrusion technology is continuously improving so this problem is being gradually reduced as new machines come to market.

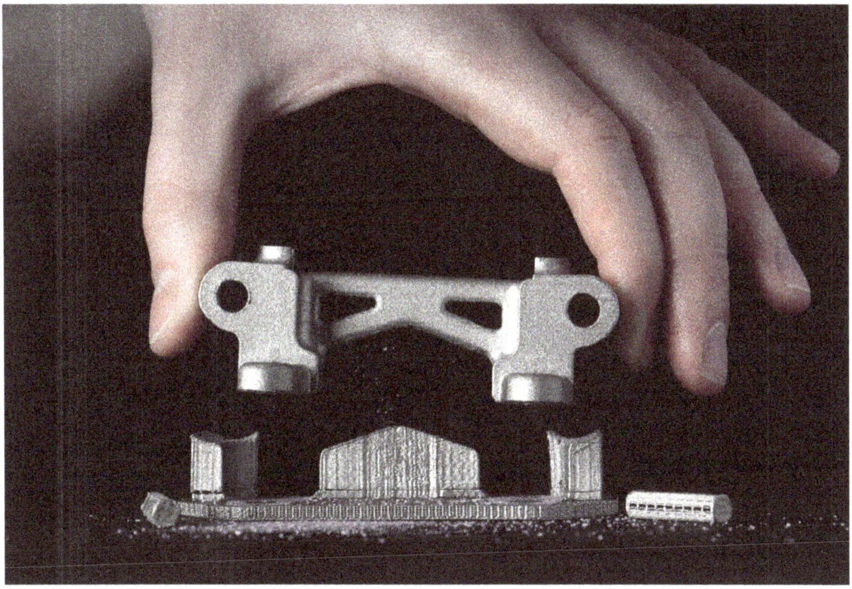

Fig. 2.2 Easy support material removal of Desktop Metal material extrusion system. Courtesy of Desktop Metal

The surface finish of material extrusion technologies can also be an issue, depending on the intended application, as it generally has the lowest quality surface finish of the technologies described in this chapter. In particular, gently sloping surface tend to have quite a visible "stair-step" or effect. **Note that this applies to all AM technologies, but is particularly visible in material extrusion technologies** (Fig. 2.3).

Typical applications of material extrusion technologies include:

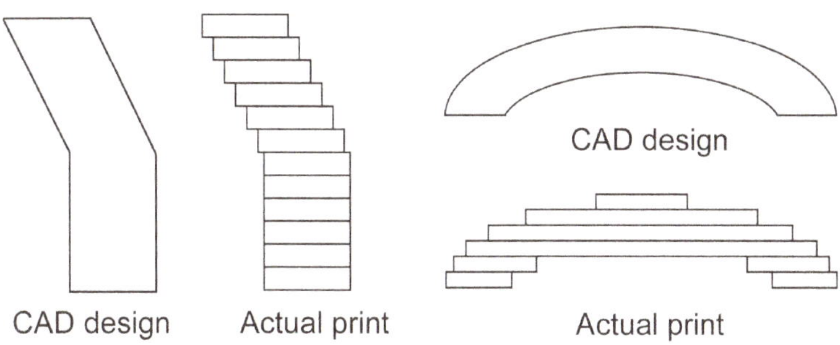

Fig. 2.3 Material extrusion stair-step effect

For prototyping

- Most applications, but the parts will require substantial post-processing for cosmetic prototypes.
- Low-cost desktop 3D printers, in particular, are great for quickly testing ideas.

For production

- Jigs and Fixtures.
- Application in which the part are not under tension in the direction of the anisotropy.
- When particular polymers are required that are not available with other technologies.
- Applications where anisotropy and surface finish are not an issue.

Material extrusion pros and cons

Pros	Cons
Most affordable machines, particularly with the advent of desktop machines (though desktop machines are not generally considered suitable for manufacturing)	Most anisotropic process. Substantial weakness in Z direction
Prints in standard engineering thermoplastics	Poorest surface quality process
Low cost material available for desktop 3D printers	Requires support material for overhangs
Easy to use machines	Potentially difficult polymer support material removal, unless they are soluble

Material extrusion materials

Standard materials	Specialty materials
ABS/ASA	Clay filled polymer
Polycarbonate	Brick filled polymer
ABS/Polycarbonate Blends	Wood filled polymer
Nylon	Metal filled polymer
PPSF/PPSU	Concrete
ULTEM 9085 and 1010	Chocolate
PLA	Polyurethane foam
Metal filled polymer filament (bronze, steel, stainless steel, copper, Inconel and others)	Silicone
	Epoxy
	Bio-materials
	HPA/PCL

2.2 Material Jetting

Material jetting technologies use a print-head, similar to that of an inkjet printer, to deposit liquid photopolymer material drop-by-drop for each slice of the model, and have a UV light-source attached to the print-head, which cures the deposited droplets of material as it passes over them. The print-head is designed to deposit both the part material, as well as any required support material, for each slice of the model.

Material jetting printers produce high-resolution parts out of a number of different plastic-like and elastomeric materials, as well as in wax, which is commonly used for investment casting and jewellery making. The newer generation of machines can also use multiple materials on the same part which makes them useful for producing full-colour parts and parts such as plastic components over-molded with rubber, for example.

The UV light that follows the print-head fully cures the materials so, unlike vat photopolymerisation (SLA), post-curing is not required. The support material does, however, need to be manually removed, typically, with a waterjet. With some technologies, though fully cured, the parts can be somewhat 'sticky' to the touch when they come off the machine, but leaving them in ambient sunlight for a day, or so, will harden the outer surface and 'dry' them off (Fig. 2.4).

The two main factors to be considered when using material jetting technologies are the post-processing steps required to remove the support material, and the fact that material jetting technologies tend to be affected by UV light, including the UV

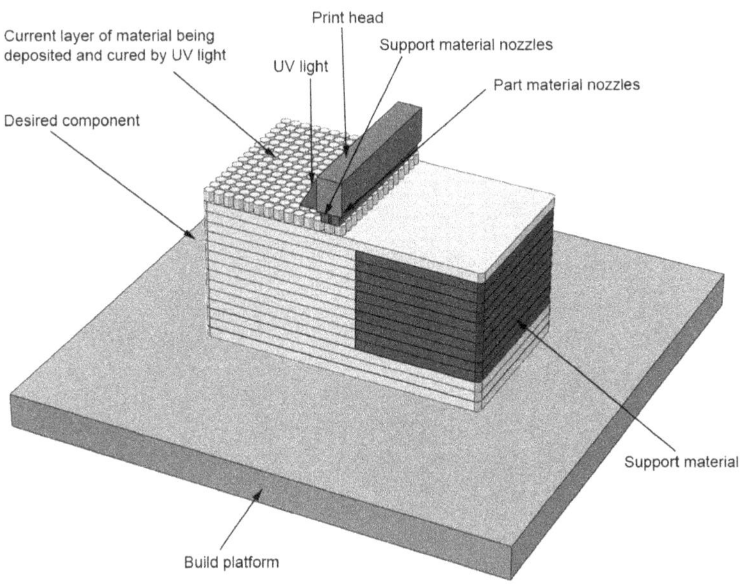

Fig. 2.4 Material jetting polymer additive manufacturing system

component of normal ambient sunlight, so are not always ideal if the unprotected component is to be used in direct sunlight, for example. If left unprotected and exposed to ambient light, the material properties, and the part colour, will change over a period of time.

Some metal material jetting systems are now appearing on the market in which a binder filled with metal or ceramic nano-particles are printed and, as the part is being printed, heat is used to evaporate the binder and fuse the metal or ceramic particles together. So far, these systems have mostly only been able to produce relatively small parts because of the shrinkage that occurs during the process (Fig. 2.5).

Typical applications of material jetting technologies include:

For prototyping

- Most applications, and one of the favourite technologies for cosmetic prototypes (together with vat photopolymerisztion) because of the high level of detail and surface quality.
- The advent of full-colour and multi-material material jetting technologies has made it a favourite technologies for cosmetic prototyping.

For production

- Applications where the product will not be exposed to ambient UV for extended periods (Hearing aids, for example).
- Investment casting patterns.
- For short-run injection molding tools.

Fig. 2.5 Metal part created by material jetting. Courtesy of Xjet

Material jetting pros and cons

Pros	Cons
Best surface finish of any AM technology (together with vat photopolymerisation)	Material properties change over time
Can make clear parts	Elastomeric materials are weak under tension
Can make multi-material and full-colour parts	Resin is messy and can be hazardous when removing parts from machine and when post-processing
Can make wax parts for investment casting	Requires support material for overhangs

Material jetting materials

Standard materials
Digital ABS polymer
High-temperature transparent polymer
Rigid opaque polymer
Simulated polypropylene polymer
Rubber-like polymer
Wax
Bio-compatible polymers
Dental materials
Metals

2.3 Binder Jetting

Binder jetting, which was also the first technology to coin the term 3D printing, is a powder-based AM technology that was developed at MIT around 1993. It was then commercialized by Z Corp, which was later acquired by 3D Systems.

The technology is one in which a layer of powder is deposited on a build-platform, and liquid binder is applied to the appropriate parts of the layer using an inkjet printer head. Wherever the binder is applied to the powder, the powder solidifies. The build-platform is then lowered, another layer of powder is spread on over the first, and the process is repeated until the part is completed within the powder bed (Fig. 2.6).

After completion, the build-platform is raised and the loose powder that has not had any binder applied to it is brushed or vacuumed away to fully expose the completed "green" part. Options are then available to strengthen components, or make them more flexible by infiltrating additives into the part.

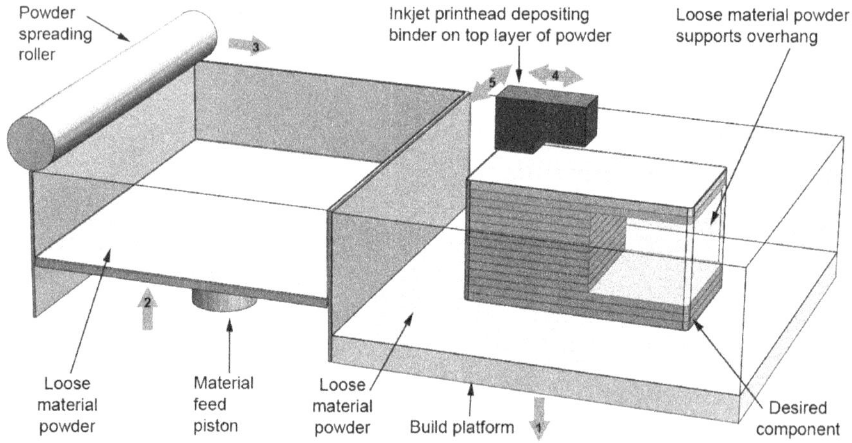

Fig. 2.6 Binder jetting additive manufacturing system

One of the advantages of this technology are that some of the materials that are available, such as gypsum powders, metal powders, PMMA and a number of sand-casting sands make it useful for producing, for example, sand molds or investment casting patterns that can be used for casting metal components in the traditional way.

Binder jetting can be further subdivided into the following areas:

Direct Part Production Binder Jetting These machines, most commonly, produce parts out of gypsum, starch, or PMMA (acrylic). The parts produced in gypsum or starch are fragile when they come out of the machine so, typically, need to be reinforced by infiltrating them with cyanoacrylate (superglue) or epoxy. Ceramic slip-casting powders can also be used in these systems to produce ceramic parts, but these parts need to be fired in a ceramic furnace after printing.

On some printers, parts produced in gypsum or PMMA can be printed in full colour. PMMA parts can also be used as patterns for investment casting.

Metal Binder Jetting These systems use metal powder that is bound together by the printed binder to produce relatively fragile 'green' parts. The category can be further broken down into two sub-categories:

Infiltrated binder jetting, in which the green parts are placed in a furnace and infiltrated with bronze, which burns the binder away, to produce alloy parts.
Full sinter binder jetting, in which the parts are sintered in a furnace as a post-processing operation, and the metal particles are fused together while the binder is burned away.

It should be noted that some of the materials can shrink by between 3 and 20% upon sintering, so this needs to be taken into account in the design. Metal binder jetting design guidelines are further discussed in a later chapter.

Sand-Casting Binder Jetting These binder jetting systems use standard sand-casting sands to produce molds, or mold cores, that can be used for the sand-casting production of metal parts.

From an industrial use point of view, this is one of the easiest technologies to use, because sand-casting has a long history of use, so engineers entirely trust the sand-cast parts. It is only the molds for the parts that are produced with additive manufacturing. The two leading companies in producing AM systems for printing sand molds are Voxeljet and ExOne (Fig. 2.7).

Typical applications of binder jetting technologies include:

For prototyping

- If using gypsum powder as the materials, in most applications where part strength is not an issue, and where colour printing adds value.
- In PMMA, and other plastic powders, the technology is suitable for most prototyping applications, and some machines can produce full colour models.

For Production

- For manufacturing sand-molds for casting.
- For manufacturing PMMA molds for investment casting.
- For direct metal part production.

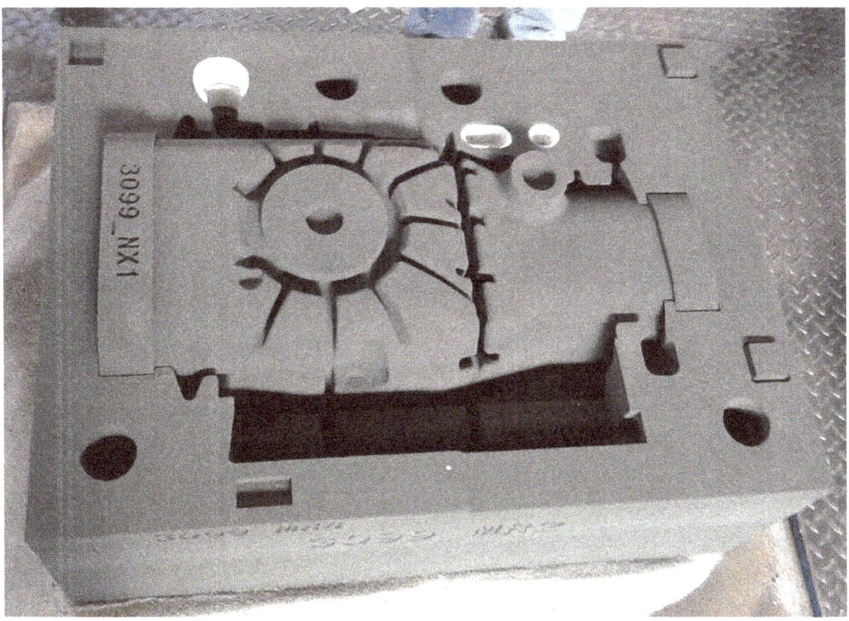

Fig. 2.7 Binder jetting printed sand-casting mold for Imperia GP Roadster gear box. Courtesy of Voxeljet AG

Binder jetting pros and cons

Pros	Cons
If printing in gypsum or PMMA powder the process can produce full colour parts	If printing in gypsum, the parts are very fragile, and will require infiltration of a strengthening resin
Can make metal parts (but will require sintering as a secondary operation)	Powders that require sintering after the green part is printed may shrink by up to 20%
Can create sand molds and cores for conventional metal casting	
Can produce investment casting patterns in PMMA	

Binder jetting materials

Standard materials	Speciality materials
Gypsum (plaster of Paris)	Sugar
Starch	Glass
PMMA	Hydroxyapatite (HPA)
Metal powders	
Sand-casting sands	
Ceramics	

2.4 Sheet Lamination

Sheet lamination processes include laminated object manufacturing (LOM) and ultrasonic additive manufacturing (UAM). Laminated object manufacturing uses paper or polymer film, which is cut out into the appropriate shape for each slice of the model using a blade, and adhesive to bond them together layer upon layer.

Laminated objects are often used for aesthetic and visual models and are not, generally, suitable for structural use. The main manufacturer, today, of LOM systems is Mcor, and their Iris and Arke systems can produce models in full colour (Fig. 2.8).

With LOM, as paper is used, the process does not require any specialist tools and is time efficient. Whilst the structural quality of parts is limited, adding adhesive, paint and sanding can improve the appearance, as well as further machining.

The Ultrasonic Additive Manufacturing (UAM) process uses sheets of metal, which are bound together using ultrasonic welding. The process does require additional CNC machining of the unbound metal. Unlike LOM, the metal cannot be easily removed by hand and unwanted material must be removed by machining. Material saving metallic tape of 0.150 mm thick and 25 mm wide does, however, result in

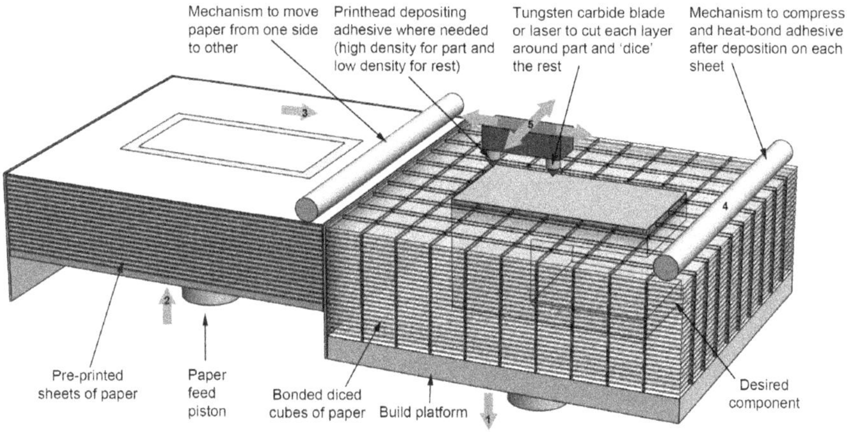

Fig. 2.8 Sheet lamination additive manufacturing system

less material to machine off afterwards. Milling can happen either after each layer is added or after the entire process.

Metals used include aluminium, copper, stainless steel and titanium. The process occurs at a low temperature and allows for internal geometries to be created. One key advantage is that the process can bond different materials and requires relatively little energy, as the metal is not melted, but instead using a combination of ultrasonic frequency and pressure to bond the layers together. Materials are bonded by plastic deformation of the metals.

Post-processing requires the extraction of the part from the surrounding sheet material. This is most commonly done with the excess material surrounding the part being machined off after each layer is bonded to the previous one.

Typical applications of sheet lamination technologies include:

For prototyping

- All aesthetic prototyping applications where low-material cost and colour adds value.

For Production

- LOM is used for producing sand casting patterns.
- UAM can be used for metal production parts, but will require post-machining.

Sheet lamination pros and cons

Pros	Cons
Benefits include speed, low cost, ease of material handling, but the strength and integrity of models is reliant on the adhesive used	Finishes can vary depending on paper or plastic material but may require post-processing to achieve desired effect
Cutting can be very fast due to the cutting route only being that of the shape outline, not the entire cross sectional area	Fusion processes require more research to further advance the process into a more mainstream positioning
	Limited material use

Sheet lamination materials

Standard materials
Paper
Plastic film
Metal

2.5 Vat Photopolymerisation

Vat photopolymerisation technologies produce components out of liquid resins that are cured (hardened) by UV light. The main technologies in this category are stereolithography (SLA) using either lasers or DLPs (Digital Light Processing), and Continuous Liquid Interface Production (CLIP).

Vat photopolymerisation uses a beam of UV laser light, or DLP with a UV light source, to cure layers of resin for each slice of the model. The beam of UV light scans the surface of the liquid photopolymer resin for each slice of the model, and wherever the beam hits the surface of the resin, it gets hardened. The build platform then descends into the resin by a fraction of a millimetre, a fresh layer of resin is spread over the top of the previous layer, and the laser scanning process for the next slice of the component starts, which both builds the slice while, at the same time, bonding it to the previous layer. The process is repeated, with the component gradually descending into the resin while fresh layers are added to the top, until the part is built. Note that some systems produce parts from the bottom up rather than top down. In these, the laser, or DLP, is placed beneath the build tank and traces or projects the slice through a clear window, and pulls the part out of the resin as it is built layer by layer.

Vat photopolymerisation produces high resolution parts out of plastic-like materials, and some of the recent advances in resin quality produce parts with enough optical clarity that, after polishing, it is suitable for printing clear components such as electronic product viewing windows and even lenses.

It should be noted that Vat photopolymerisation produces support material scaffolding to support any unsupported overhanging parts of the component. After the build is complete, this support material needs to be manually removed, and the part then needs to be post-cured in a UV oven to fully harden the material (Fig. 2.9).

As with material jetting technologies, the parts are susceptible to degradation if left unprotected and exposed to ambient UV light.

Ceramic Vat Photopolymerisation Several companies, including Lithoz, 3D Ceram, Prodway and Admatec now offer photopolymer resins filled with ceramic powder that allow for the production of high-density ceramic parts. Examples of some of the available ceramic materials include Aluminiumoxide (Al_2O_3), Zirconiumoxide (ZrO_2), Tricalciumphosphate ($Ca_3(PO_4)_2$).

These ceramic vat photopolymerisation processes produce green parts that need to be sintered in a furnace as a secondary post-processing operation. Shrinkage of between 15% to 30% occurs during this sintering process, so careful design consideration needs to be given to this. The process is, today, mainly used for producing relatively small ceramic parts.

Continuous Liquid Interface Production (CLIP) is a vat photopolymerisation process from Carbon. The main difference between this process and the other bottom-up vat photopolymerisation processes lies in the oxygen permeable membrane that

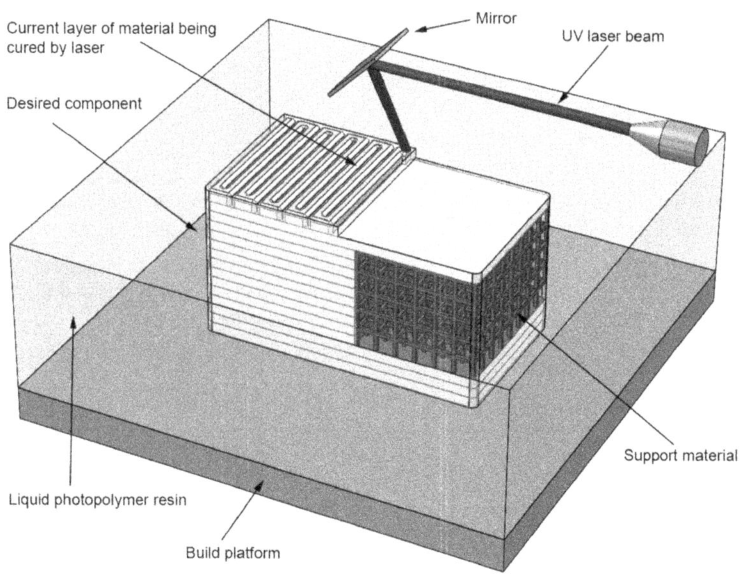

Fig. 2.9 Vat photopolymerisation additive manufacturing system

the UV light shines through. With the other processes, when the resin hardens, it sticks to the bottom window, and has to be delaminated though mechanical force before the resin can flow into the gap. But, with CLIP, the hardened resin does not stick to the window and, for thin parts, the resin can easily flow into the gap.

When first released, CLIP gained much notoriety through advertising claims of being from 25 to 100 times faster than other technologies. However, with parts that are not made of thin lattice-like structures, the speed advantage largely disappears, because the resin cannot flow quickly into the gap between the part and the window when the surface area is large.

An advantage of CLIP technology is that it uses polyurethane resins, so it is one of the few vat photopolymerisation that can make parts that do not rapidly change properties over time. However, it requires a secondary process, of heating up the part in an oven to cross-link the polymer to create the polyurethane. This time required for the secondary curing process is also often overlooked when doing time comparison calculations.

Typical applications of vat photopolymerisation technologies include:

For prototyping

- Most applications, and one of the favourite technologies for cosmetic prototypes (together with material jetting) because of the high level of detail and surface quality.

For Production

- Applications where the product will not be exposed to ambient UV for extended periods (Hearing aids, for example).
- Investment casting patterns.

Vat photopolymerisation pros and cons

Pros	Cons
Best surface finish of any AM technology (together with material jetting)	Material properties change over time
Can make clear parts	Resin is messy and can be hazardous when removing parts from machine and when post-processing
	Requires support material for overhangs

Vat photopolymerisation materials

Standard materials
UV Curable photopolymer resin
A number of ceramic filled resins
Polyurethanes for CLIP

2.6 Powder Bed Fusion

Powder bed fusion technologies, in the context of rapid manufacturing, include Laser Sintering, Selective Laser Melting (LS/SLM) and Electron Beam Melting (EBM). All these technologies work by spreading a fine layer of the build material, in powder form, across the build platform, and then using a beam of energy (a laser in the case of LS and SLM, and an electron beam in the case of EBM) to scan the slice of the component and melt the powder wherever the laser hits the powder. The build platform is then dropped a fraction of a millimetre, and the next layer of powder is spread across the platform, and the melting process repeated, both melting the current layer and, at the same time, bonding it to the previous layer.

These powder technologies have recently advanced to the point of being able to produce full strength components that are relatively isotropic in their behaviour in the X, Y and Z directions. There can be some anisotropy in the Z direction but, if the part is well designed for AM, this can be minimized, or eliminated through post-processing. In addition, these technologies can produce parts in both a variety of polymers (in the case of LS) and metals (in the case of SLM and EBM). Note that there exist several other acronyms for metal powder bed fusion technologies, and these are listed in the acronyms chapter of this book.

Typical polymers include a range of Polyamide plastics (Nylon) with a number of fillers including glass, carbon fibre, aluminium, and also high temperature polymers such as PEEK. Plastic components, if well designed for AM, exhibit properties similar to injection molded components, and include the ability to make living-hinges (if designed for AM) and clip together components. The surface finish, straight out of the machine, is that of a matt plastic component, and some 'stair-step' effect is visible on gently sloping or curved components.

Metals include stainless steel, aluminium, titanium, cobalt chrome, maraging steel (tool steel), and more. Metal parts are generally comparable, both in surface finish and in strength to a cast component without any voids, occlusions or other defects. The implication of this is that, much like any cast component, if completely smooth, or polished, surfaces are required, then they will need to be achieved through post-processing on a CNC machine (Fig. 2.10).

Multi jet fusion, a relative new-comer on the AM scene, deserves a special mention as it differs somewhat from the other powder bed fusion processes. With multi jet fusion, which derived from selective heat sintering (SHS), a fusing agent, and a detailing agent, is inkjet-printed onto the layer of polymer powder, and the entire print bed is heated with an infrared heat source. Wherever the fusing agent has been printed, the powder absorbs enough of the heat energy to melt, whereas the rest of the material remains in unfused powder form (Fig. 2.11).

Typical applications of powder bed fusion technologies include:

For prototyping

- Most applications, from aesthetic to functional prototypes.

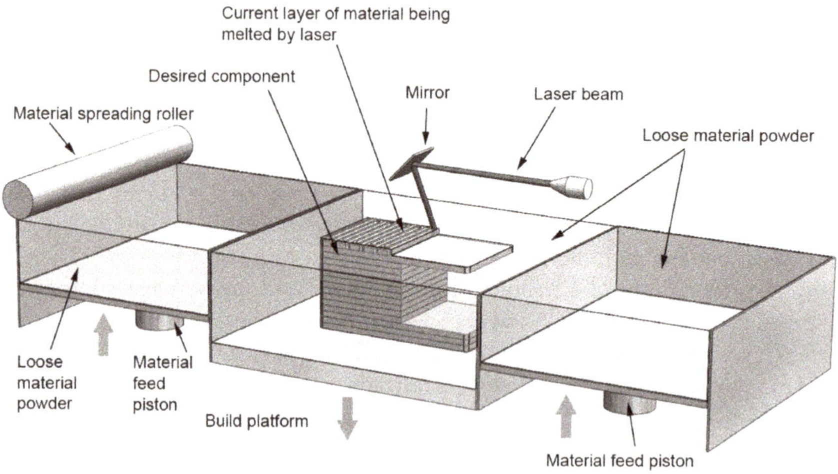

Fig. 2.10 Powder bed fusion additive manufacturing system

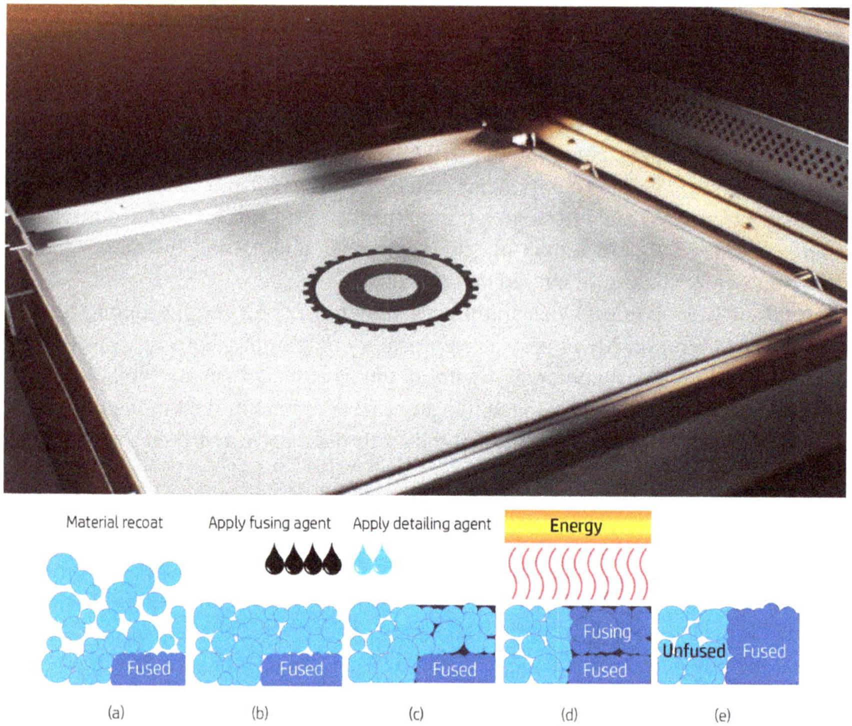

Fig. 2.11 Multi jet fusion additive manufacturing system. Courtesy of HP

For Production

- Powder bed fusion, both polymer and metal, is the most used technology to produce industrial production parts.
- Polystyrene material is also used for investment casting patterns.

Powder bed fusion pros and cons

Pros	Cons
From a material point of view, one of the lowest cost production technologies	Metal PBF requires support material for heat transfer, and can require considerable effort to remove
Produces strong and durable parts	
No support material required for polymer powder bed fusion	

Powder bed fusion materials

Polymer materials	Metal materials
Nylon 12, 11 and 6	Stainless steel
Glass filled nylon	Maraging steel (tool steel)
Heat resistant nylon	Titanium 64
Polypropylene-like nylon	Aluminum
Alumide (aluminium-filled nylon)	Tungsten
Carbonmide (carbon-filled nylon)	Nickel-based super alloys
PEEK	Cobalt chrome
	Copper
	Precious metals, such as gold

2.7 Direct Energy Deposition

Directed Energy Deposition (DED) covers a range of technologies including LENS (Laser engineered net shaping), DMD (direct metal deposition), CLAD (3D laser cladding). It is a near net-shape printing process commonly used to repair or add additional material to existing components. Though all AM processes can be considered "near net-shape", DED produces considerably less accurate parts, and a poorer surface finish, than most other AM technologies so, in the majority of cases, will require a machining post-processing operation.

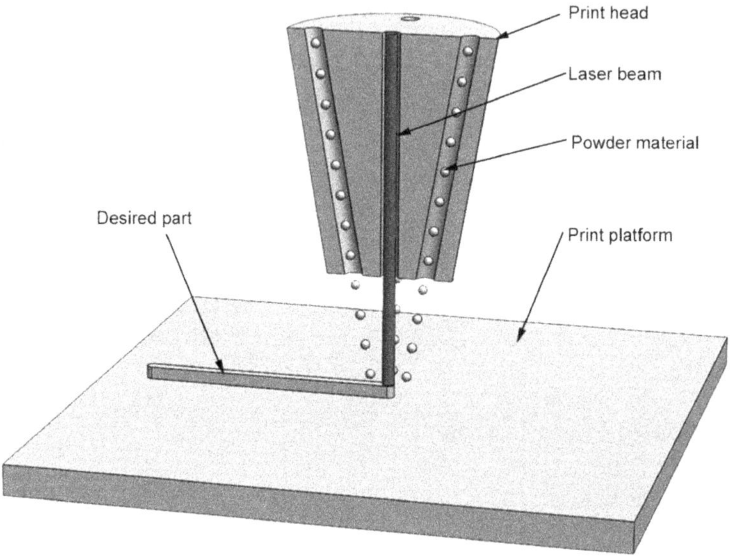

Fig. 2.12 Directed energy deposition additive manufacturing system

A typical DED machine consists of a nozzle mounted on a multi axis arm, which deposits powder or wire material onto the specified surface and into the path of an energy beam, where it melts and solidifies. The process is similar in principle to material extrusion, but the nozzle can move in multiple directions and is not fixed to a specific axis. The material, which can be deposited from any angle due to 4 and 5 axis machines, is melted upon deposition with a laser or electron beam. The process is typically used with metals, in the form of either powder or wire (Fig. 2.12).

Typical applications of directed energy deposition technologies include:

For prototyping

- Not generally used for prototyping.

For Production

- Applications where a near net shape is acceptable, that will be machined as a secondary operation.
- Can be used for repairing damaged or worn parts.

Directed energy deposition pros and cons

Pros	Cons
Fast process	Near net-shape surface finish that, in most cases, will require machining
5 axis process so can be used to repair worn or damaged parts without needing to machine the surface flat	

Directed energy deposition materials

Standard materials
Almost any metal that is weldable, and available in powder or wire form

2.8 Hybrid AM

Hybrid additive manufacturing systems that combine powder or wire based DED AM and CNC machining in a single system are now available.

The great advantage of hybrid systems is that they are one of the few technologies that can produce ready-to-use parts straight off the machine (excluding heat treatment). The disadvantages include the still unknown factors in the metallurgical properties of the parts, as they are subjected to cooling (during machining) and heating (during AM) in the process.

The machines also require a more experienced operator that has the knowledge of operating 5 axis CNC machine and a DED AM system. The choice about when to use the systems AM features and the CNC features is still, largely, a manual process that the operator has to make decisions about. As the technology develops, automated software will, no doubt, be developed that makes the use of hybrid AM technologies easier to use (Fig. 2.13).

It should be noted that those technologies briefly described above are only a few of the currently available additive manufacturing technologies. The reasons we cover these, and not the others, is that these are the most widely adopted by industry today, and are the most applicable to rapid manufacturing applications. New AM technologies are, however, developing at a fast rate and one would expect a large increase in uptake of these technologies, for manufacturing purposes, over the upcoming years.

Fig. 2.13 Hybrid directed energy deposition CNC additive manufacturing system. Courtesy of DMG MORI

2.9 AM Technology Readiness Level for Part Production

All of the AM technologies described above are suitable for prototyping applications. It is, however, important for the design engineer to gain an understanding of the strengths and weaknesses of each technology in order to understand to what level each particular technology is suitable for producing production quality parts. Some will argue that every technology can produce full strength production quality parts, while others will dispute this claim.

For metal production parts, we would classify the order of the technology readiness level of various AM technologies as follows:

For direct part production

1. Powder bed fusion, for small to medium sized parts.
2. Direct energy deposition, for larger parts or part repair, but will require machining.
3. Full sinter binder jetting, for precise small parts, or medium parts that require less precision.
4. Hybrid AM, for medium to large parts.

For indirect part production

1. Binder jetting of sand-casting molds.
2. Vat photomolymerisation and material jetting for investment casting patterns.
3. All technologies for sand casting patterns.

For polymer part production, we would classify the technology readiness levels as follows:

For direct part production

1. Powder bed fusion.
2. Material extrusion for jigs and fixtures.
3. If a part is not subjected to ambient UV then Vat photomolymerisation and material jetting.

For indirect part production

1. Powder bed fusion and material extrusion for vacuum forming molds.
2. Vat photomolymerisation and material jetting for short-run injection molds.

Chapter 3
DfAM Strategic Design Considerations

3.1 Introduction to Design for Additive Manufacturing

Design for additive manufacturing (DfAM) is when designers seek to create a product design that takes advantage of the unique capabilities of AM. DfAM also respects the specific process constraints of the AM technology that will be used to produce the product. This goes beyond merely re-designing existing parts for AM. Re-design for AM is useful because it can yield benefits such as a reduction in the use of material or the consolidation of several parts into one. However, what it fails to do is to consider the added benefits that AM can bring to an entire product through improvements in form, fit, and function. This book seeks to encourage engineers and designers to consider the strategic benefits of AM before concentrating on detailed design. Design for AM is definitely more of a thought process in which conscious decisions are made (often compromises) rather than just blindly following a set of design rules.

As interest in additive manufacturing keeps growing for manufacturers around the world, it is becoming more important to, whenever possible, make sure that parts are specifically designed for AM. There are, of course, some application areas, such as spare parts, in which previously conventionally made components will not get redesigned for AM. In general, however, there are often things that can easily be done, even to conventional parts, to minimise the print times and costs.

In this book we talk about any form of the redesign of parts to better suit the AM process as 'design for AM'. A more distinct hierarchy of the different types of design processes used in the additive manufacturing spectrum might, however, distinguish between only modifying the AM production process parameters, modifying the form of the part, but not its function to better suit the AM process, and entirely redesigning the form and function of the part to be truly designed for AM. Typically, these three different approaches are referred to as direct part replacement, adapt for AM, and design for AM (Fig. 3.1).

An example of this is shown in the various design options for the manifold shown in Fig. 3.2.

O. Diegel et al., *A Practical Guide to Design for Additive Manufacturing*, Springer Series in Advanced Manufacturing,
https://doi.org/10.1007/978-981-13-8281-9_3

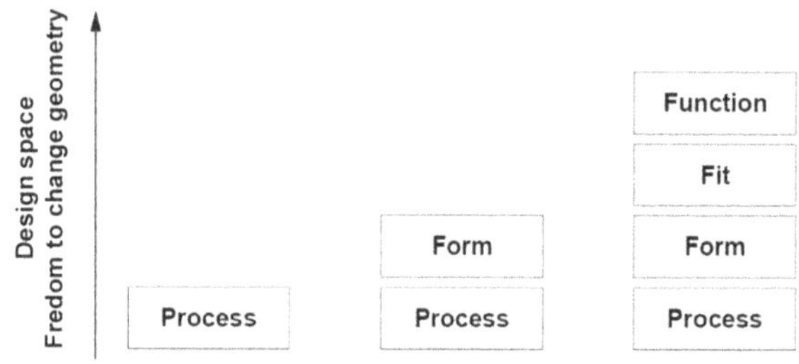

Direct part replacement	Adapt for AM	Design for AM
This is used when absolutely no change to the part is allowed, and the part must be reproduced as closely as possible to the original part. The main reason for using this approach is in spare parts when lead-time is important enough to justify the use of AM	Changes are made to the form of the part, often internally, to make the part easier to manufacture through AM. The external shape of the part might also be changed, but it's use and function, and how it fits into the product does not	The entire part is redesigned to maximize the benefits of AM, and for how the part will be printed. Here we reconsider how the part fits in within its surrounding product and what and how it performs it function and one attempts to improve this

Fig. 3.1 Distinguishing between AM in direct part replacement, adapt for AM, and design for AM

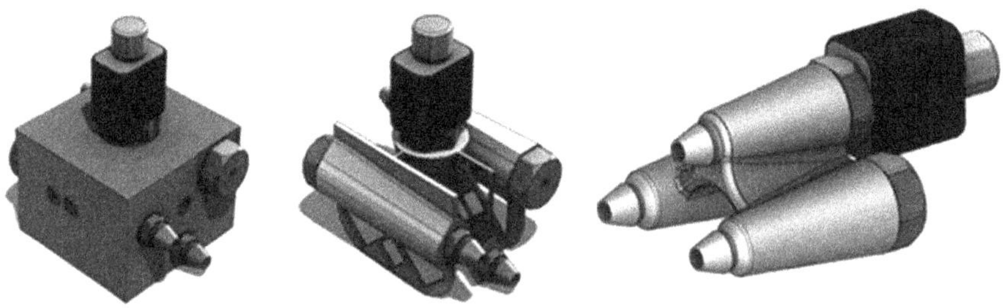

Direct part replacement	Adapt for Am	Design for AM
4.6 kg	1.0 kg 78% weight saving.	0.4 kg 91% weight saving. Improved fluid flow and fit within product. Less space required in product, and easier assembly.

Fig. 3.2 Design approaches for manifold, courtesy of Marc Saunders, Renishaw

3.2 Using AM to Add Value to Products

As stated earlier, AM is an expensive manufacturing technology and, because of its 'serial' production method it will always be relatively slow in comparison to some conventional manufacturing technologies, so is likely to remain comparatively expensive even in the future. Therefore, when considering AM for production parts, in the vast majority of cases, it should only be used if it is truly adding value to the product.

Before starting individual part design, engineers and designers should analyse their product to see which value-adding strategies they should adopt. This will have an impact on the product structure and part configuration. After this step, they can begin to design for specific AM processes, the subject covered in the following chapters of this book.

3.3 General Guidelines for Designing AM Parts

There are a number of fundamental principles that can be applied to almost any form of additive manufacturing. These are described below, and most will be further discussed in subsequent chapters.

3.3.1 The #1 Rule of Design for AM

It Depends!

There are few rules of design for AM that can universally be applied to all geometries, materials, AM technologies and parts. Many design parameters depend on other design parameters and printing conditions, so precise numbers that work in every case are hard to find. But, as an AM engineer or designer gains a deeper understanding and experience of the intricacies of AM processes, they gradually become better equipped to design parts that will print perfectly the first time.

An example of the variability of design guidelines is the minimum hole or slot size through a powder bed fusion part, which depends on the thickness of the part. As the part grows in thickness, unfused powder in narrow holes gets partially fused in the holes, so becomes impossible to remove. But different powder bed fusion machines also produce parts of different qualities, as they run at different temperatures, layer thicknesses, and laser scan parameters. So the minimum hole or slot size is directly related to the thickness of the part, the layer thickness, print orientation, as well as to the machine it is made on (Fig. 3.3).

The same goes with the clearances between moving parts. The larger the surface area of the parts that are in close contact, the larger the gap between the moving parts

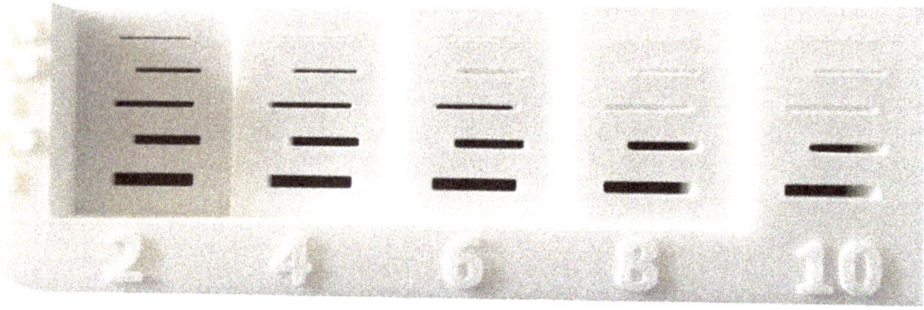

Fig. 3.3 Hole size depends on material thickness

must be. This is, again, because of the larger surface areas retaining heat for longer and causing powder between the moving parts to fuse.

Many of the numbers given in the section of this book on design guidelines for specific AM processes should only be used as a general guidelines, as other parameters of the part design may affect the given numbers. When in doubt, it is always best to print a test part to ensure that the numbers will work for that particular context.

3.3.2 The #2 Rule of Design for AM

Should you be using AM in the first place?

There are countless examples of the use of AM to produce parts, when the part could be produced much better, cheaper, and faster using a number of other technologies. This is further described in the economics of additive manufacturing chapter. As a general guideline, if a part was designed to be machined on a 3-axis CNC machine, it will be cheaper and faster to CNC machine it than to 3D print it. Though there are, of course, exceptions to this, the general guideline is, nonetheless, true.

When used for production, in particular, AM should only be used if the part cannot be easily be made using some other manufacturing technology. The principle reason for this is speed, which translates to cost. Comparatively speaking, AM is relatively slow compared to almost all other manufacturing technologies. Therefore, from a cost point of view, it will almost always be more economical to manufacture geometrically simple parts using traditional technologies if they are faster than AM. However, when the complexity of a part reaches a point where it cannot be manufactured conventionally, that's when AM starts to shine (Fig. 3.4).

In the above example, the simple parts on the left of the red line can easily be made through laser cutting, waterjet cutting, punching, CNC machining, etc. The more complex parts on the right of the red line, however, would be hard to manufacture conventionally, so are ideal for AM.

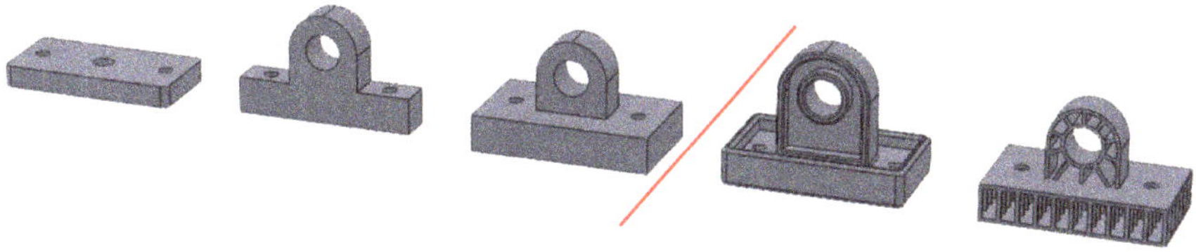

Fig. 3.4 Complexity selection filter

3.3.3 The #3 Rule of Design for AM

It doesn't cost any more to make things beautiful!
The design freedom characteristics of AM allow you to make almost any shape you can imagine. If possible, use this to give your product a unique aesthetic. It also doesn't cost anything more to add useful cosmetic details, logos, instructions, part numbers, etc. onto your parts. This can help make assembly easier, help to identify your product brand, and keep track of stock more easily.

3.3.4 The #4 Rule of Design for AM

Fillet all corners.
It can be good practice to fillet (round) all sharp edges. This serves a dual purpose: First, it makes the product more ergonomic and comfortable to hold and use, as it removes the risks of sharp edges, and second, it reduces the stress concentrations that can occur on sharp corners and transitions which could affect the strength of the product.

Internal corners, in particular, are where stress concentrations will occur so, if there is no reason not to, internal corners should always be filleted. But, as it costs no more to print sharp external corners than rounded external corners (in fact, it costs a bit less as they require less material to be melted), it is also good practice to fillet external corners.

A good rule of thumb is to make the fillet ¼ of the thickness.

3.3.5 The #5 Rule of Design for AM

You cannot design for AM without thinking of print orientation.
The quality (strength, material properties, surface quality, amount of support, etc.) of every part, with every AM process, is directly related to print orientation. When

Fig. 3.5 Some of the effects of printing a part in two different orientations

you design, you should therefore continuously be thinking about which orientation the part will be produced in.

Print orientation determines the direction of the anisotropy, which will always be in the Z direction, or vertical print direction. So, if anisotropy is an important factor, the part should be oriented so as to have the features requiring maximum strength to be printed horizontally.

If the roundness of holes is also critical, they are best printed in the vertical orientation. Holes printed horizontally will suffer from the stair-step effect, and will also be slightly elliptical.

The total height of the build determines how many layers of material it will require, and therefore how long it will take to print, which will affect the cost. If there are no other critical considerations, the best print orientation is usually the one that minimizes the total height of the build.

Figure 3.5 shows some of the differences in part quality when printed in different orientations on a polymer powder bed fusion system.

3.3.6 The #6 Rule of Design for AM

Design to minimize large masses of material.
Large masses of material in a part cost a lot, cause a lot of residual stress, and add little engineering value. Any feature that breaks the 'even-thickness rule' for no good reason is just unnecessary material that increases cost, causes more residual stress and therefore requires more support material and heat treatment.

When CNC machining, for example, we leave large masses of material, to avoid having to machine them away because this would cost time and money. AM is the opposite way around, because any unnecessary material adds to the print time and cost, and should therefore be avoided.

3.3.7 *The #7 Rule of Design for AM*

Design to minimize support material.
Print orientation, described in rule #6, also determines where support material will be required for overhangs, and for heat transfer. Support material location is, therefore, ALWAYS a key design consideration.

$$\textbf{Support material} = \textbf{Labour} = \textbf{Cost}$$

$$\textbf{Support material} = \textbf{Part quality}$$

Most AM technologies require the use of support material. As the placement and amount of support material greatly affect part quality and post-processing cost, its use must be carefully taken into account during the design process. Both part design, and print orientation affect the amount, and position of support material. If support material is placed in the wrong areas, it can be very difficult and time-consuming to remove. The next chapter, on the economics of AM further explains the reasons for attempting to reduce the amount of support material used. Support material is also a recurring theme throughout this book.

3.4 Design to Avoid Anisotropy

Design for AM rule #5 is about making decisions about print orientation. One of the chief reasons for this is because of anisotropy. Anisotropy, the difference in mechanical properties of a part in the vertical direction, can be the Achilles heel of additive manufacturing. This anisotropy affects all AM technologies, but some are affected more than others (Fig. 3.6).

The left bracket would be weak in the direction of the force, whereas the right bracket would be substantially stronger because of the print orientation.

Technology	Effect of anisotropy in vertical (z) direction
Material extrusion	Greatest
Photo-polymerization	Medium
Polymer powder-bed fusion	Medium to minimal depending on feature thickness
Metal powder-bed fusion and other metal technologies	Least, and can be mostly eliminated through heat treatment or other post-processing such as hot-isostatic pressing (HIP)

When designing your component, it is good to have an understanding of the orientation that it will be built in, so that it can be designed with the optimal number

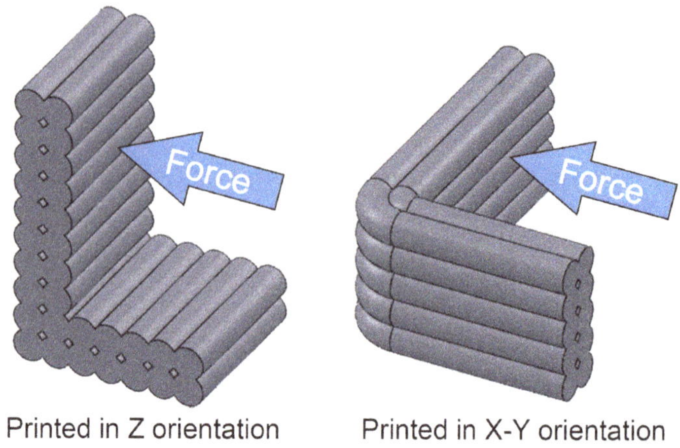

Fig. 3.6 Anisotropy, or the weakness between layers, will cause apart to react differently under different forces

of features that will be subjected to high forces oriented in the X-Y plane. This is discussed further in the chapters on designing for polymer and metal AM processes.

3.5 The Economics of Additive Manufacturing

Some of the hype around the last few decades of AM has been that AM can produce lower-cost parts. Though, in some case, this is true it is, in general, an expensive and slow technology that can only be used for production if it adds enough value to the product beyond that which could be added with conventional manufacturing.

The graph is often used to demonstrate how, for conventional manufacturing, costs decrease as quantities increase, whereas with AM, costs remain relatively constant. Many people misconstrue the red line in the graph as indicating that AM is cheaper than conventional manufacturing and this is, in fact, quite rarely the case (Fig. 3.7).

Many industries approach additive manufacturing with the idea that they can simply use it to replace the conventional manufacturing technologies they currently use, without re-designing their components. This is, almost always, a mistake as none of the benefits offered by AM are being exploited. A production component that was designed for 3-axis CNC machining, for example, will generally be more expensive to manufacture with AM, and is likely to still require CNC machining to give it the required surface quality that cannot be obtained with AM. There is general agreement that, for AM to offer maximum value, the parts must be designed for additive manufacturing (DfAM).

So why is designing for AM so important? There are many reasons for this, including taking advantage of the functional and aesthetic value that "complexity for free" adds. The most compelling reason, however, is simple economics. This chapter

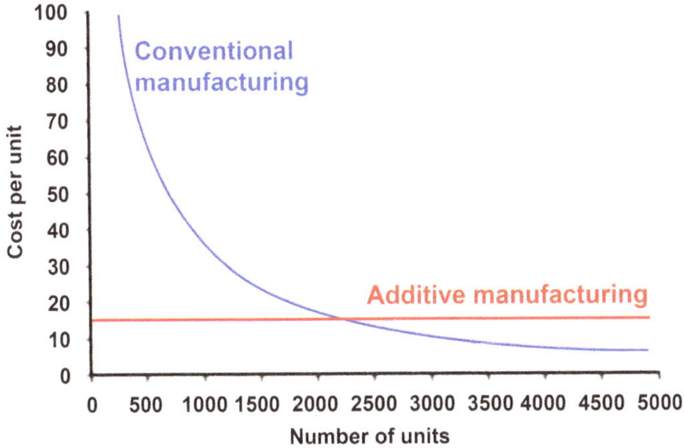

Fig. 3.7 Graph showing how costs decrease as quantity increases with conventional manufacturing but remain constant with AM

explores some of the economic arguments that, for metal AM in particular, make DfAM a necessity rather than a luxury. Although metal AM is used in this section as an example, as it is the most susceptible to print time and post-processing activities, the contents of this chapter applies to all AM technologies.

Additive manufacturing, for components or tools that have not been specifically designed for AM, can be very expensive to manufacture in metal. The reasons for this are relatively straight forward: Metal AM systems are expensive, and part production rates are slow. A metal production sized AM system, typically, costs between US$500,000 to over US$1,500,000.

One can, optimistically, assume that a metal machine will be running approximately 80% of the time, so about 7000 h per year. The reasons for only operating 80% of the time include maintenance, machine cleaning and preparation time, heating up or purging time, cool down time, part removal time, etc.

A common payback period used by industry to recoup the cost of an infrastructure investment is 2 years. Some companies, of course, use a longer payback period, but as an AM machine is in an area of technology that is evolving rapidly, 2 years is not an uncommon payback period. On top of this, there may be a loan for the investment in the machine for which there may be an interest rate of, say, 5%. There are also a few other costs involved, such as energy and labour costs to setup the machine, but these are relatively minor in comparison to the machine cost time.

Machine hourly running costs = (*Machine purchase cost* + *interest*)/(*payback period* × % *running time* × *yearly hours*)

Machine purchase cost	Hourly machine running cost
$500,000	$37.45/h
$650,000	$48.69/h
$1,000,000	$74.91/h
$1,200,000	$89.89/h

This demonstrates that the typical hourly operation costs of a metal AM machine can range from about $37/h to $90/h (depending on the value of the machine). For the sake of the calculations in this chapter, let's pick a mid-range operating cost of $65/h. Note that the above is an extremely simplified calculation, and could also include many other factors including overheads, and other factors.

This means that, if a single part takes 10 h to print, the machine cost for that part is $650. With metal AM, however, particularly when parts have not been optimised for AM, print times often range to substantially more than this, and it is not uncommon to have print times of 40, 60, or even well over 100 h. For a single part with a 100 h print time, for example, the machine cost of the single part is $6500.

Note that a good CNC machine can be in a similar price range as a metal AM system, thus having comparable hourly operating costs. The difference is that a typical 'simple' part can be CNC machined in a much shorter time than it can be made on an AM system. Compared to an injection molding machine in a similar price range, the difference is even more extreme. The time to injection mold a part, typically, ranges from a fraction of a second to a few seconds.

One of the other costs associated with AM part production is the raw material. For aluminium and steel powders the prices are, typically, around $30/kg to $90/kg, while other alloys, such as cobalt chrome and titanium can rang up to $300/kg. With metal AM, parts require support material to support overhangs and, more importantly, to anchor the part to the build platform and conduct heat away from the top layer. This typically means that there is about 10% material wastage (including both support material, and partially sintered powder particles that get wasted in the sieving process). Though this material cost must, of course, be taken into account, in comparison to the above machine costs, it is relatively small.

On top of the direct machine costs, with metal AM there are often substantial post-processing costs. This cost includes the time it takes heat-treat the part, remove the part from the build platform, remove all the support material from the part, and then give it an acceptable surface finish.

In the 2018 Wohlers Report, service providers were asked, in 2017, what percentage of their part costs were attributed to printing versus pre- and post-processing. The results were as shown in the table below. Some companies, including Daimler, Premium Aerotec, EOS and Materialise estimate that as much as 70% of part cost can be spent on pre- and post-processing.

	Metal (%)	Polymer (%)	Both (%)
Pre-processing	13.2	10.9	10.0
Post-processing	31.4	20.2	27.0
Total pre and post	*44.6*	*31.1*	*37.0*
Printing	55.4	68.3	63.0

Even with the conservative estimate that 45% of the part costs is in pre- and post-processing, the cost for the above 100 h part jumps up to $12,000. Now designing a part to both minimize print time, and minimize post-processing time suddenly takes on increased significance!

The next two sections describe ways of minimizing print and post-processing time, and are focused on metal AM, as this is the technique in which these factors have the most impact. The same principles and thought process will, however, also apply to most other AM processes.

3.5.1 Time Factors That Are Not Affected by Design

There are some factors in metal AM part production that take time, and therefore cost, that are not affected by the part design. Recoater time is an example of this. Recoater time is the time it takes for the AM system to spread a layer of powder before the laser can start to melt the layer. Typical recoater times are around 4–15 s per layer depending on the machine used. Using an average of, say, 8 s, if a part is 100 mm tall, and the layer thickness is 50 μs, the part will contain 2000 layers, and the total recoater times will therefore be 16,000 s, or about 4.5 h which, using the average machine hourly price above of $65/h, equates to about $290 of recoater time.

Recoater time, however, is not a factor that is affected by design (other than by reducing the part height), as a well-designed part and a badly-designed part of the same height will have the same number of recoater layers.

The same applies to machine purge times. Metal AM machines manufacture parts in an inert gas atmosphere (usually argon, nitrogen, or a vacuum), and the purge time is the time it takes to remove the oxygen from the build chamber. This can take between 10 min to around 2 h depending on the machine. Some machines also require heating of the build chamber or the build plate, and this too takes a certain amount of time. Again, the time it takes for these two factors costs time/money, but is not affected by good or poor design.

The table below shows the various steps involved in metal AM, and the steps for which the total print times are affected by the design of the part.

AM process step	Affected by design
Pre-processing	
• Check quality of files, and fix if necessary	No
• Prepare print-job in software by arranging parts on build platform	No
Printing	
• Clean the AM system	No
• Purge the system of oxygen	No
• Preheat the AM system	No
• Print the parts	
– Spread layer of powder (Recoater time)	No
– *Laser contour lines*	*Yes*
– *Laser hatch pattern*	*Yes*
• Remove build platform from machine	No
• Recycle powder	No
Post-processing	
• *Thermal stress relief*	*Yes*
• Remove parts from build plate	No
• Hot Isostatic Pressing (HIP)	No
• *Removal of support structures*	*Yes*
• *Heat treatment*	*Yes*
• Surface machining, shot-peening, abrasive flow machining, etc.	No
• Inspection	No

3.6 Design to Minimize Print Time

The main factor, that is controllable through design, that determines the time it will take to print a metal part is the amount of powder that needs to be melted, or deposited, in each layer of the part. Most metal AM systems work by having the material melted in a 'serial' fashion in which the energy beam 'draws' each slice of the model on the powder to melt it. In additive manufacturing, this is referred to as contour lines and hatch patterns.

This is the same principle as sketching a filled square with a pencil. We first draw the outer edge of the square (the contour line), and then scan (or hatch) the pencil back and forth several hundred times to fill in the square. This means that, the more surface area there is to fill, the longer the distance the energy beam has to travel, and the longer time it takes to create each slice of the model (Fig. 3.8).

Take the hydraulic manifold shown below as an example. It is designed to be manufactured with CNC machining, and consists of a metal block into which a number of holes are drilled, forming interconnecting channels through which hydraulic fluid can flow to the appropriate ports.

If one were to make the above manifold using AM, any particular slice of the manifold would look like a filled square with a few holes in it, as shown in Fig. 3.9.

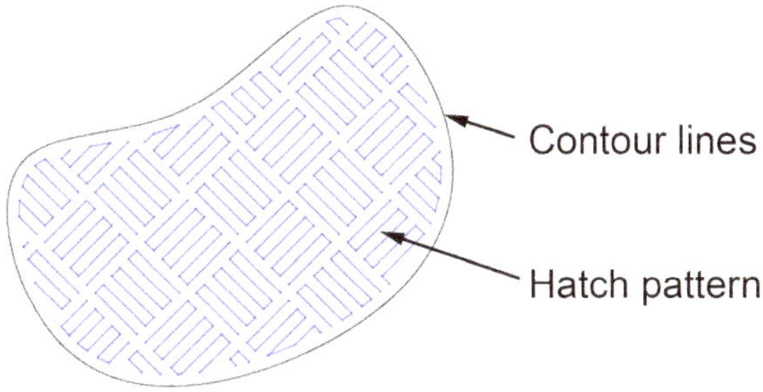

Fig. 3.8 The contour lines and hatch patterns that many AM systems use to solidify or deposit material

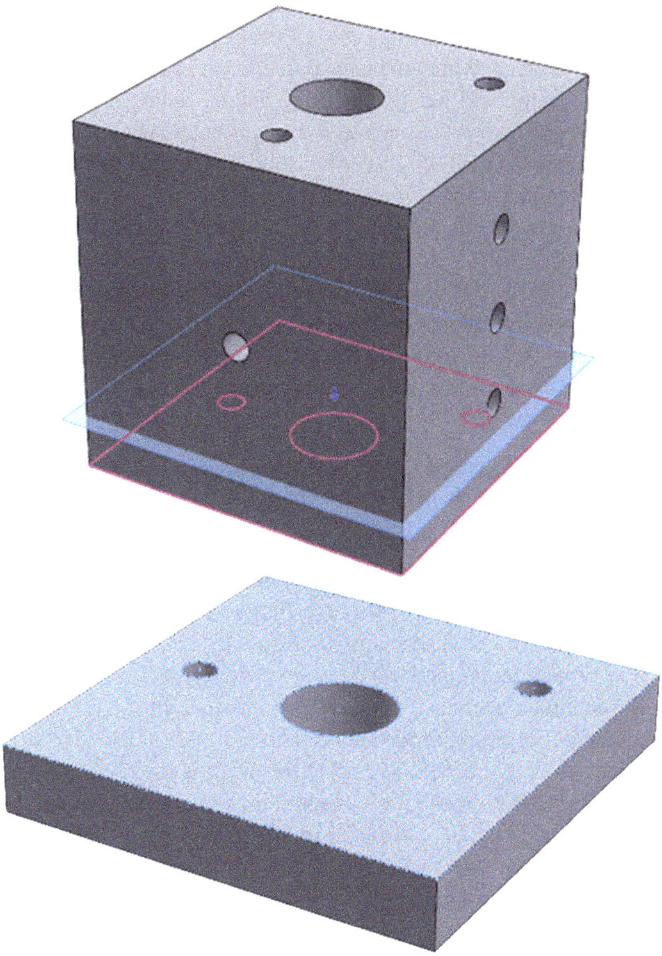

Fig. 3.9 Scanning pattern for slice of solid designed manifold

The scanning pattern for the above slice contains many meters of laser scanning distance. If the manifold measures 100 mm × 100 mm, and the hatch spacing is set to, say, 0.1 mm, then each square will require approximately 100 m of scanning to cover both the contour lines and hatching. In other words, the laser has to travel over 100 m to create that slice of the model. This takes time, which equals money. If the laser is travelling at 330 mm/s, it will take 300 s, or 5 min, to hatch that slice of the model, or $5.41 of machine time per slice, using our average $65/h machine cost.

$5.41 per slice may not sound too bad but, if the layer thickness is, say, 50 μs (which would be considered a relatively thick layer thickness), then 2000 layers will be required to make the part. So just He laser scanning time for the part is $10,820.

Shelling to reduce print time.
In contrast, if the same part is "shelled" (which means the bulk of the material is removed from inside the part, leaving only a specified wall thickness), the scanning distance is greatly reduced, which means a much faster print time. If the shell thickness is set to, say, 2 mm, and we use the same hatch spacing parameters as above, then the total scan distance is only about 4.5 m for each slice of the part. That's a scan distance reduction of over 95% over the solid model and means the laser has to do 95% less work for each slice of the model. If the laser is travelling at 330 mm/s, it will now only take 13.6 s to hatch that slice of the model, or $0.24 of machine time per slice (Fig. 3.10).

When the part is finished, the internal cavities will be filled with powder, which can be left there if weight is not an issue. If weight is a concern, then salt-shaker holes can be added so that the internal powder can be removed. The part interior will also be filled with support structures, but they can be left in the manifold without affecting the application.

It is worth reinforcing that above stated rule of DfAM because it is so important: One of the most important rules of design for AM is to try and avoid large masses of material.

Beyond the cost implications, large masses of material, typically, offer little engineering advantage and, in fact, can be detrimental as those are the areas where high residual stresses will be found, and that are likely to cause distortion in the part. And, as shown above, in additive manufacturing they also greatly extend the print times. There are many techniques for removing large masses of material, including the 'shell' technique described above. Others include filling the solid parts with honey-comb, lattices, or even porous material.

In this respect, the logic of additive manufacturing is the reverse of that used in subtractive manufacturing. With CNC machining, for example, we always aim to minimize the amount of cutting (material removal) that has to happen in order to minimize time, so we leave a lot of material behind. With AM, the less material we use in our design, the faster the part can be made, and the lower the costs.

It is very important to understand that the scan time for each layer is not the only factor that affects print time. Other factors include recoater time (the time it takes to spread a new layer of powder on top of the previous one), and layer pre-heat time, but most of these can only be minimized by reducing the build height, rather than

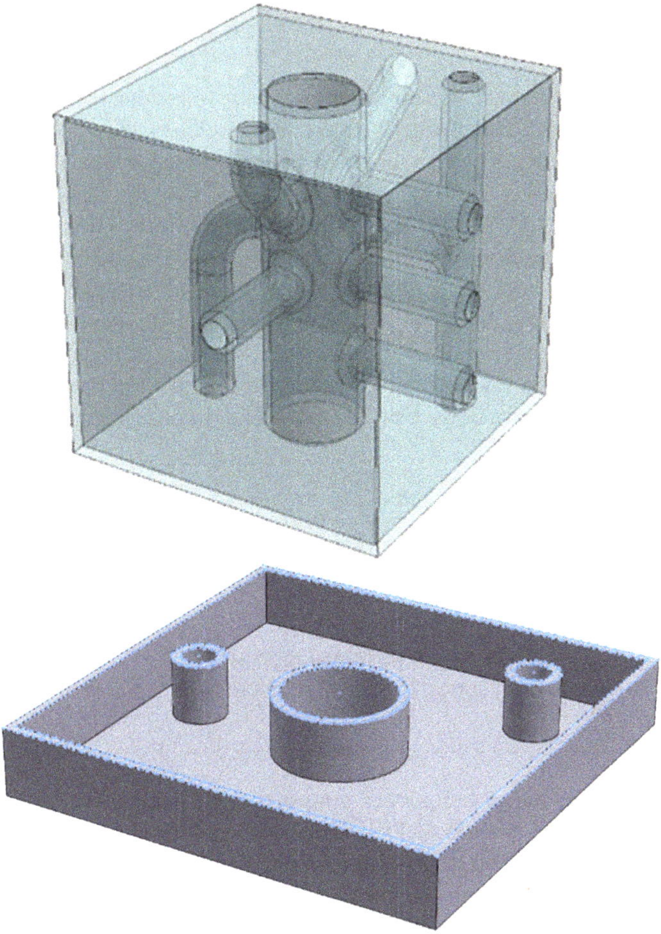

Fig. 3.10 Scanning pattern for slice of shell designed manifold

the design of the part. The bigger, in the vertical direction, the part is to build, the more layers it will contain, so the more recoater and layer preheat times it will need, so the longer it will take to print.

From a pure time perspective, it is best to print a part in the orientation in which it has the lowest vertical height as this will, generally, have the fastest print time. But, because print orientation also plays a major role on part mechanical properties, geometry accuracy, surface finish, and support material, this decision can often be a compromise of print time versus mechanical and geometry properties versus surface quality versus support material removal.

The support material minimization example, in the next section, further demonstrates how material removal can be beneficial for additive manufacturing.

3.7 Design to Minimize Post-processing

One of the important goals in designing for AM is to reduce the amount of support material used when printing the part. As mentioned above, support material with metal AM is used to anchor the component, to help support overhanging features and to transfer heat away from the component. Typically, any feature that is more than a certain angle from vertical (this angle varies depending on which material is being printed) will require support material. In some cases using a wall as a feature of the design can be used to both avoid this support material while, at the same time, improving the strength of the part. In particular, it is important to avoid support material in any internal features, for example, inside the channels in the manifold described above, as this can be difficult, or impossible, to remove.

The overall thought process that can be useful in redesigning parts for AM includes:

1. Reduce part to only those features that serve a functionality.

Anything that breaks the 'even-thickness rule' is just unnecessary material that increases cost, causes more residual stress and therefore more supports and heat treatment.

2. Decide how those features can be joined together.
3. Now consider the most appropriate print orientation depending on what is important to you.
4. Run it through support generation software to see results.

    - **Consider replacing temporary supports with permanent walls**. Support material can be thought of as a temporary wall that will be removed after the part is printed. So why not consider replacing the temporary wall with a permanent one that then becomes a feature of the part.
    - **Consider changing the angles of features requiring support**. If a feature is horizontal, it will require support material beneath it. But, if you can change its angle, chamfer or gusset the bottom horizontal face at 45° then the need for support material can be avoided.

5. Reiterate.

Now, let us examine the overall step-by-step design thought process that can be employed in the redesign of the above 100 mm × 100 mm × 100 mm steel manifold.

The design process begins with the simplifying the original 'block' design by eliminating all the drilled holes that have been blocked by plugs (in other words, all the holes that serve no other functional purpose than to allow the internal channels to be created). What we are after is the simplest possible representation of the 'block' manifold with just the actual channels for the transport of hydraulic fluid. It can also be useful, at this stage, to smooth out some of the channel junctions with fillets to make the flow of fluid smoother than the original straight drilled holes may have allowed for (Fig. 3.11).

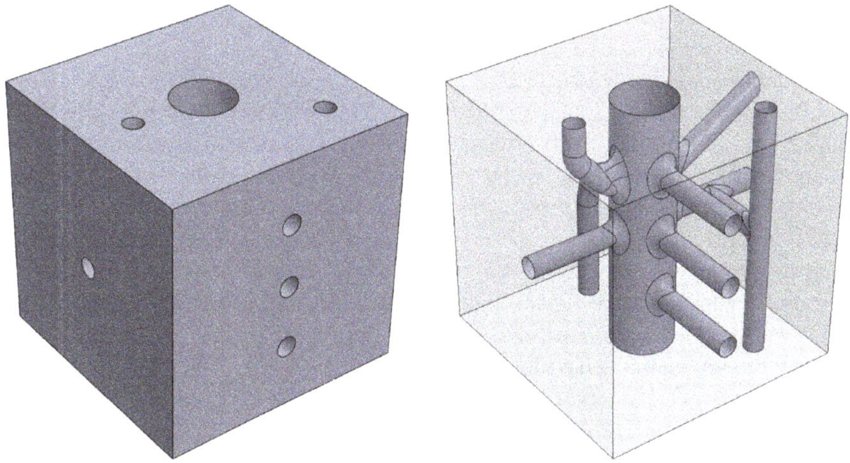

Fig. 3.11 Simplified 'block' design manifold with only the required in and out channels

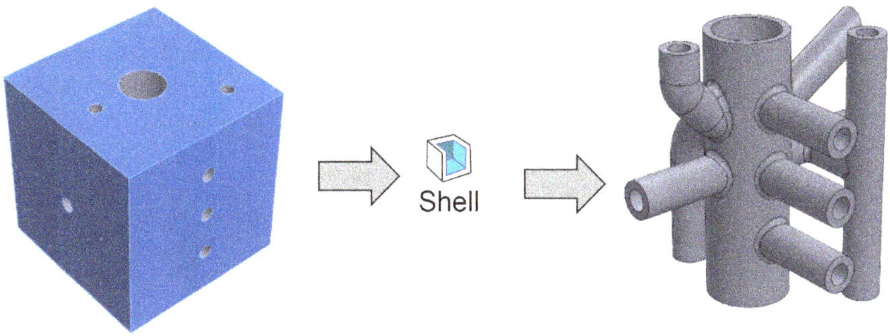

Fig. 3.12 Manifold design after shell operation on block design

Once the 'block' design has been suitably simplified, the next step is to remove all the excess material of the cube to leave just the pipes that form the manifold channels. In other words, we reduce the part to only those features that serve a functionality. Most CAD software packages have a 'shell' function that allows a component to have faces removed leaving only a shell of the product of a specified wall thickness. In the case of this exercise, we simply select all six outer faces of the cube to be removed, leaving only the internal channel structure with, in this case, a thickness of 2 mm (Fig. 3.12).

Also, now that the overall manifold can be better visualized is a good time to decide whether it's functionality could be improved by, for example, changing some of the channels that currently stick out horizontally to have a bend that makes them come out vertically. If so, the easiest is often to go back and modify the original 'block' design and then redo the shell function. In this case, let us imagine that the design is as functionally sound as it can be, so the next step is to examine the design from an additive manufacturing optimisation point of view.

An important factor to consider at this stage of the design is the print orientation, as this will affect all other design decisions. **When designing for additive manufacturing, one should always design around the specific orientation in which the part will be printed because part orientation will determine the direction of anisotropy, surface finish, roundness of holes, support material, etc.**

The first design decision we need to make is which orientation to print the part in, as different print orientations will generate support material in different places. When we run the above design through software used to generate support structures, in this case Magics, we can see that support is generated between all the horizontal pipes. In the orientation where the large diameter pipe is horizontal, we can see that support has also been generated inside the large diameter pipe (Fig. 3.13).

Both these print orientations are fine, in principle, but will necessitate the removal of this support material after printing, as well as some surface treatment to improve the surface finish of the areas where the support material makes contact with the real part. This, of course, increases the amount of labour required to finish the part, extends the delivery time on the part, and increases cost. It could also be argued that, in the print orientation where the large diameter pipe is horizontal, it will be harder to remove the support from inside the pipe than from all the outside surfaces. So, unless there were some other advantage to having the large diameter pipe printed

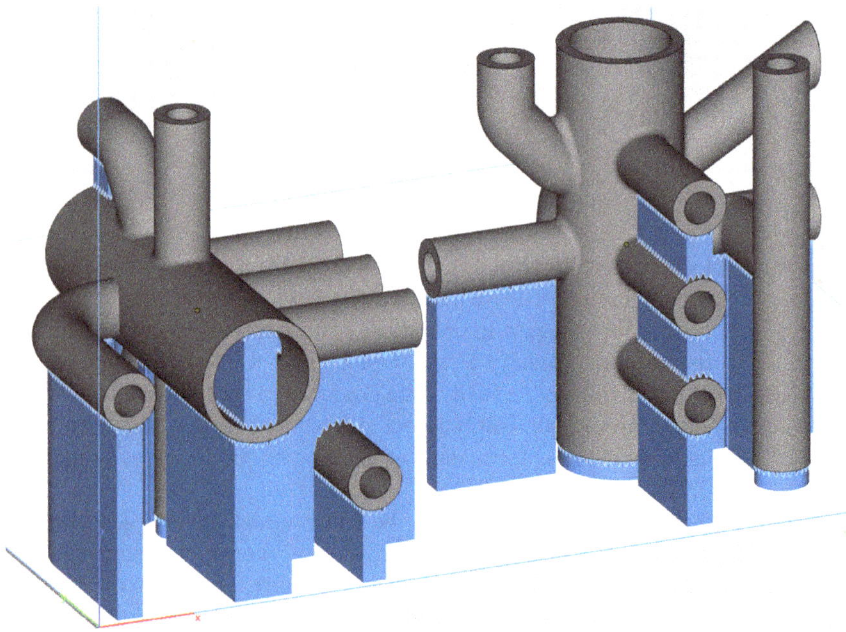

Fig. 3.13 Support material required by shelled-block design in 2 different print orientations

horizontally, the better print orientation would be the one where it is in the vertical position.

If the sole purpose of this design exercise were to achieve maximum weight reduction, then this design exercise could now be complete as it contains only the bare minimum of required material in the final part.

A design option that is worth considering, however, is to put a thin wall in beneath each of the horizontal channels to eliminate the need for support material altogether. **The idea is that the added wall becomes the support material, and becomes a permanent feature of the part.**

In the case of the example below, the bottom walls are chamfered at 45°, as that is the angle we have set beyond which to use support material, and we have added oval holes in the walls to reduce the weight, yet without requiring support material in the holes. As can be seen, with this new design, the only needed support material is that required to weld the part to the build platform.

There can also be a functional benefit to adding such support walls, as it makes the pipe substantially more rigid. As one can imagine, when one is screwing the hydraulic fittings into the manifold, a spanner may be used to tighten them on, and that can apply substantial lateral forces to the pipes. The added walls can counteract some of this force and help to minimize the risk of damage (Fig. 3.14).

In this example, when the part is removed from the build platform, through either wire EDM or sawing, the part is pretty much ready to go, other than a quick shot-peening operation, and perhaps tapping the threaded parts of the manifold.

When designing for additive manufacturing, there exist a number of design rules and guidelines dictating factors such as minimum wall thicknesses, the hole diameter above which support material will be required, the angle beyond which support

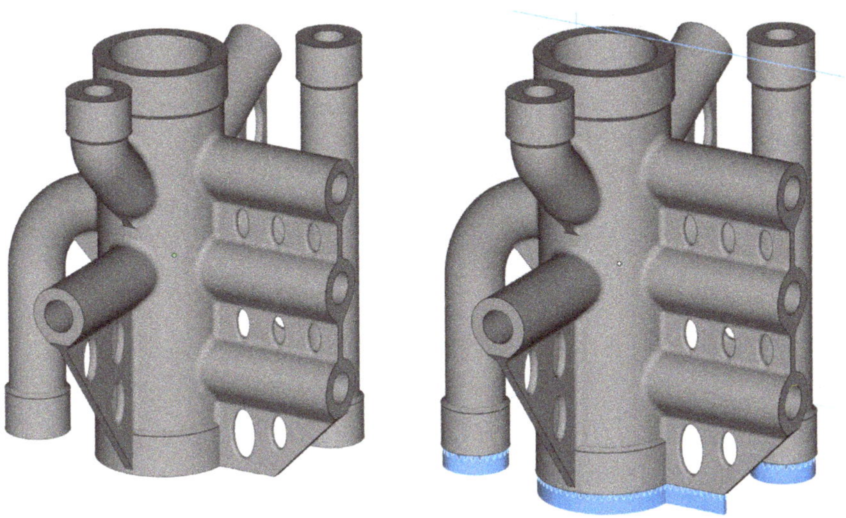

Fig. 3.14 Support material required by optimised for metal AM design

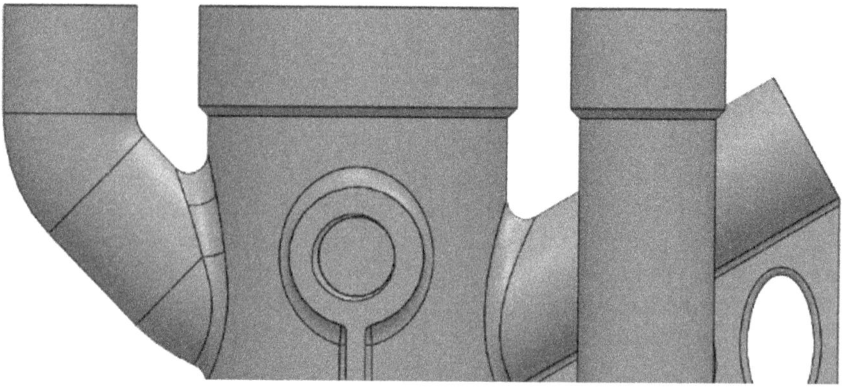

Fig. 3.15 Material added to the pipes that will require subsequent tapping, with 45 chamfer to eliminate support material

material will be required, etc. These are further discussed in the relevant later chapters of this book. In the above design example, we have carefully sized all the horizontal pipes to have an inner diameter for which support material is not required (usually between 6 and 8 mm depending on the AM system being used). We have also taken the opportunity to add a few other features, such as adding some material where pipes will be tapped as a post-processing operation. Here, again, we have optimised the design with a 45° chamfer to eliminate the need for support material (Fig. 3.15).

In the case of the above example the weight of the original 100 mm × 100 mm × 100 mm block manifold design, in steel, would be 7.4 kg. In contrast, the optimised for metal AM design weighs only 600 g. That represents a greater than 94% weight saving, not to mention a greatly reduced print time and cost.

In many cases of printing metal AM parts, some support material is unavoidable. The above example demonstrates how some simple design decisions can greatly remove the amount of support material required, and therefore, the post-processing, which means a better product at a lower cost.

If the simple 100 mm × 100 mm × 100 mm manifold design described in the section above were to be manufactured using either of the 3 proposed designs, it would cost as follows:

	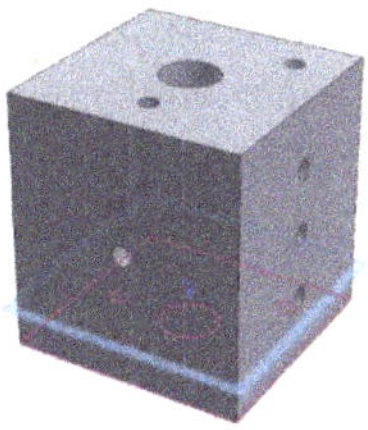	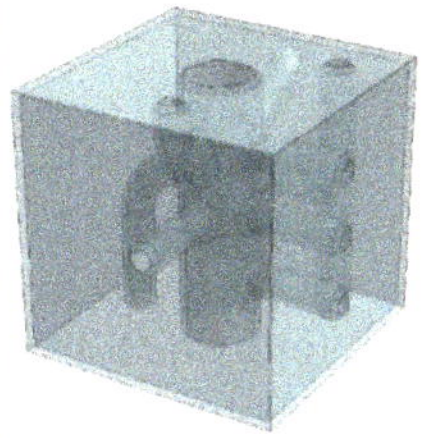	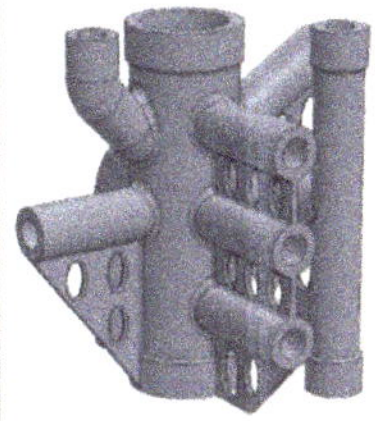
	Solid block manifold	Shelled block manifold	Optimised DfAM manifold
Scan time for hatch pattern	191 h 1 min 33 s	36 h 31 min 21 s	19 h 40 min 39 s
Machine cost in metal@$65/h	$12,415.00	$2379.00	$1261.00
Material weight	7.411 kg	1.232 kg	0.558 kg
Material cost @ $70/kg + 10% waste	$570.64	$94.86	$42.96
Bureau quotes for part in 316L stainless	$15,293.82	$3735.12	$1986.25

It demonstrates that the machine cost is one of the main factors in AM part production, and that the controllable factor that affects machine time the most are large masses of material that require extended hatching times. It quantifiably demonstrates that even a simple strategy of replacing large masses of material with even wall thickness shells can have a substantial impact on hatching times, and therefore machine time and cost. Note that we are not suggesting that shelling is the best strategy to use. Only that it clearly demonstrates the effect of avoiding large masses of material.

Using the block manifold as an example, the exercise then demonstrates some further DfAM techniques that allow the weight and hatching time to be even further decreased, as well as the technique of using real walls instead of support material in order to minimize post-processing times. Another equally effective technique is to change the angle of features so that they are always at a steeper angle than that which requires support material.

This exercise demonstrates the importance of design for additive manufacturing from an economic point of view. The difference in cost between a part that has been designed for AM compared to that of a conventional part made with AM is so significant that it shows DfAM to be an absolutely essential part of the process of making parts with AM. Add to that the less quantifiable added value elements such as improved functionality, reduced time to market, reduced waste material, and so on, and the argument for the absolute necessity of DfAM is not hard to make.

Below is a real-world example of a hydraulic manifold used on an Atlas Copco underground drilling rig which used the above techniques to reduce the manifolds weight by over 90% and greatly increase its functionality. The redesigned Atlas

Copco manifold required minimal amounts of support material. Also, the required support material was in easy-to-access locations, making it relatively straightforward to remove. After being redesigned for metal AM, the manifold's weight dropped from 14.6 to 1.3 kg—a weight reduction of more than 90%.

Beyond weight reduction, the functionality of the manifold was also improved. On the original design, the positions of the inlet and outlet ports were determined by the easiest direction from which the holes could be drilled, rather than where they would be most suitable for use or assembly. On the redesigned version, the outlet ports were moved to the top surface and only the inlet remained at the bottom surface. This greatly reduced the overall volume required to install the manifold in the machine (Figs. 3.16, 3.17 and 3.18).

Almost any product, no matter how trivial, can be treated as a good exercise in designing for metal additive manufacturing. Even something as fun, and politically incorrect, as a mini desktop distillery can make a good example of DfAM (Fig. 3.19).

The above product was treated as a design challenge to try and design the little still to use no support material at all, other than what was required to weld it to the build plate. The still measures 117 mm × 58 mm (which is the diameter of the barrels) × 66 mm high.

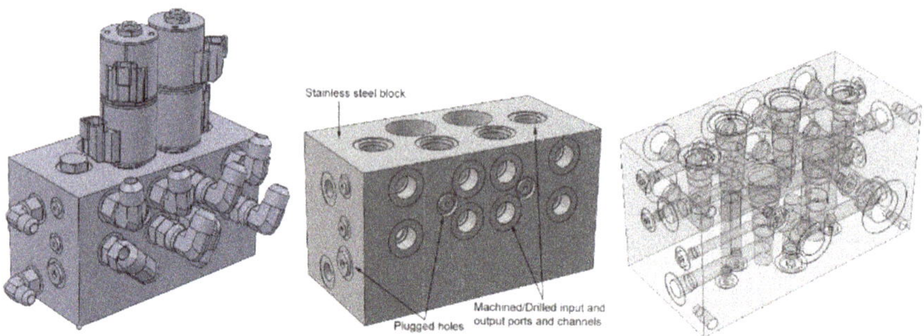

Fig. 3.16 Original Atlas Copco manifold block design

Fig. 3.17 Hydraulic manifold designed for AM with some unremoved powder (left), support structures attached (center), and support material removed (right)

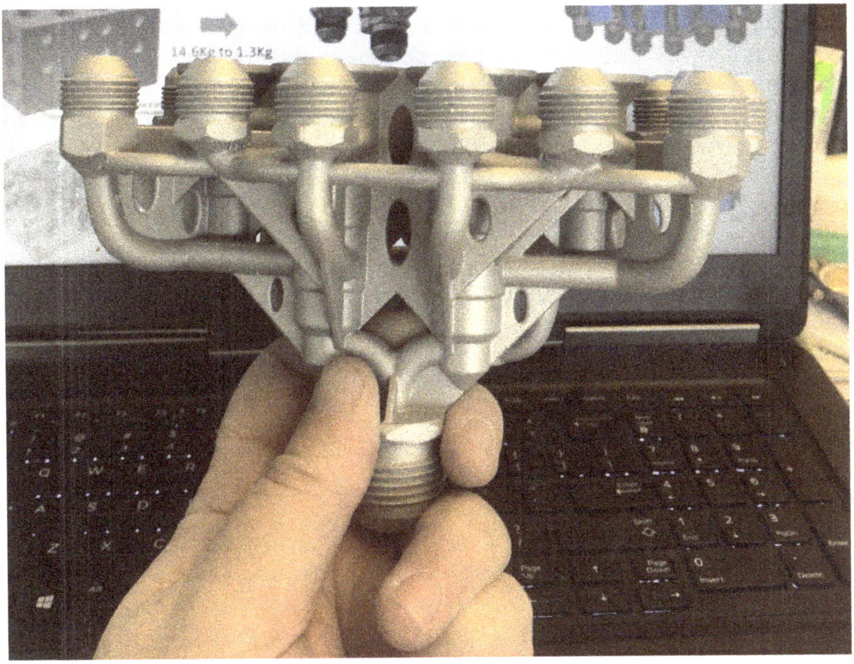

Fig. 3.18 Finished hydraulic manifold produced by metal AM

The design began by using the same "walls instead of support" technique described in the above manifold design example to support the spiral pipe that forms the condenser for the still. If this had not been done, it would have required support material between the spiral coils, which would then have been difficult to remove (Fig. 3.20).

In general, any overhangs with a surface area of more than a few square millimetres should be avoided, as well as features with angles greater than 45° from vertical, as they will then require support material. This angle can, of course, vary depending on the material being printed. In this project, the still was printed in aluminium so the design made sure that nothing exceeded an angle of 45° (Fig. 3.21).

With this design, it meant that, after printing, the still was just cut off the build plate, shot-peened, and it was ready to be used.

It is important that design for additive manufacturing be done consciously, with sound decisions about which strategies to use in order to minimize the part cost and the amount of post-processing that will be required, or about which other design for AM guidelines to apply. There is nothing at all wrong with using a bit of support material and, in fact, it is often unavoidable. But it should be done with the design engineer making a conscious design decision as to where it should be, and whether there are design strategies that can be used to avoid it without compromising the functionality of the part.

Fig. 3.19 Aluminium distillery manufactured with AM

3.8 Take Advantage of Design Complexity

The design complexity freedom allowed by AM lets you to make almost any shape you can imagine. If possible, use this advantage to give your product a unique aesthetic.

- It doesn't cost anything more to add cosmetic details, so take the opportunity to add any added-value embellishments, such as company logos, part numbers, storage locations, alignment aids or operational instructions directly on the part.
- Use shells, ribs, gussets, trusses, topology optimisation, etc., to make the product as light as possible.
- Fillet all sharp edges. This aids both in making the product more comfortable to hold and handle, as well as reducing stress concentrations that can be induced by sharp corners. Filleting all sharp edges also removes the problem that 'knife edges', particularly vertical ones can cause when building a part.

Fig. 3.20 Using walls to avoid needing support material

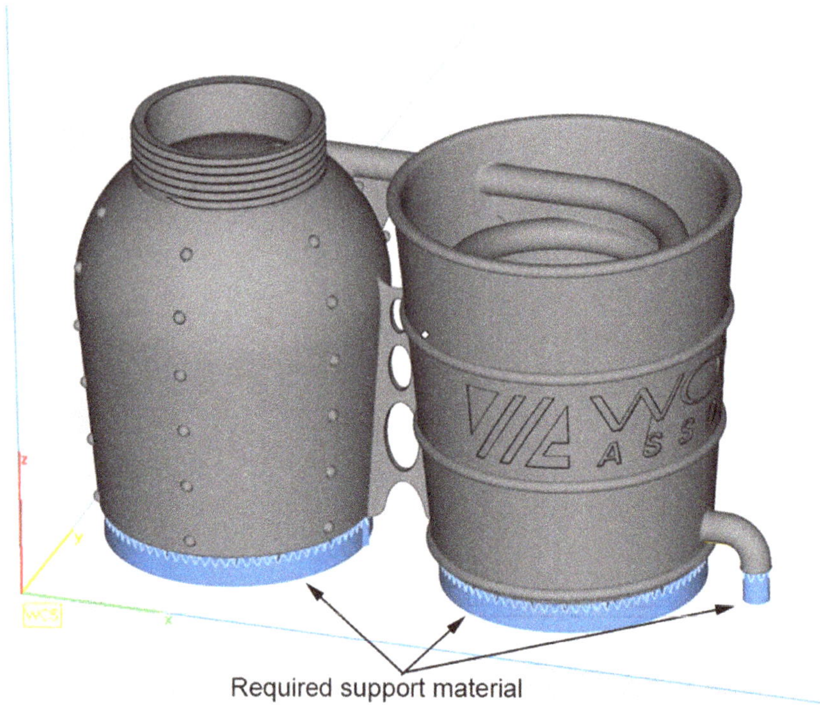

Fig. 3.21 Support material required for distillery print

3.9 Function First, Materials Second

Engineers have a tendency to focus on a material that a similar part has historically been made of, and often insist that an AM part must be made of the same material. Because of the new way in which AM lets you design, it is often best to first think of the function the part must perform and design around that. Once the design is complete, one can then look at available AM materials to see if one of them is suitable for the function, and the mechanical properties, that must be achieved.

Because a component that has been designed for AM often requires substantially less material than a conventionally designed part, it may be possible to manufacture the part using a higher-specification, more expensive, material without it costing any more than if the conventional design were made with a lower-specification material.

3.10 Use Topology Optimisation or Lattice Structures

The complexity allowed by AM means that the production of topology optimised parts is now possible. Topology optimisation (TO) is a numerical methodology that optimises material layout within a given design space and for given boundary conditions such that the resulting layout meets a prescribed set of performance targets. In other words, it uses mathematics to remove any material that is not performing a useful function within a part.

With traditional manufacturing technologies, designs from a topology optimisation, although optimal in terms of the design criteria, may be expensive or infeasible to manufacture. AM technology now, however, allows the manufacture of such complex, optimised shapes (Fig. 3.22).

The middle bottle opener would be machined from the solid aluminium block at the top, and weigh just over 4 g. In contrast the topology optimised bottle opener at the bottom weighs less than a gram.

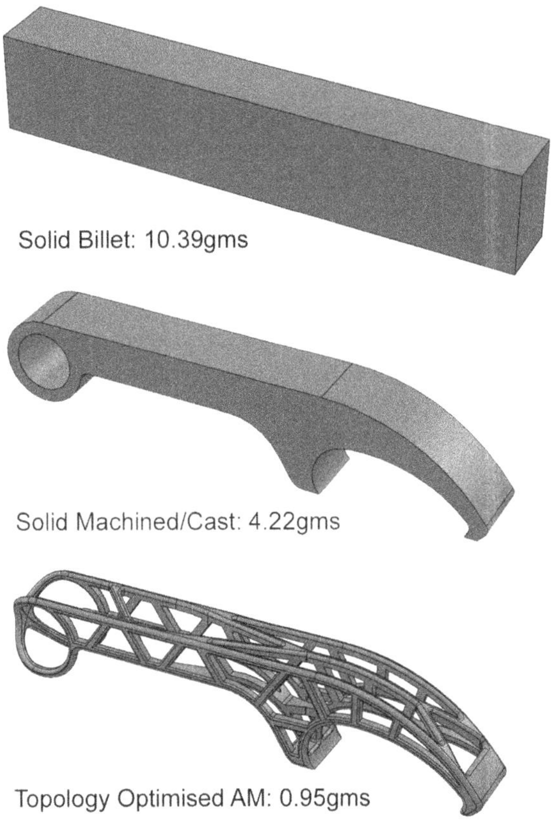

Fig. 3.22 Example of bottle opener weight reduction through topology optimisation

There are now a great many topology optimisation software packages. These include, for example:

Specialized topology optimisation software:

- Altair Optistruct and solidThinking Inspire
- Tosca
- Top3D (MatLab)
- Paramatters
- LiveParts.

Most high-end FEA products:

- Abaqus
- Nastran, etc.

The general work process flow of most topology optimisation is as follows:

1. **Simplify the model**: Remove all features from the design that were a result of traditional manufacturing. What we are looking for is a relatively large 'block' of material that the software can work with to optimise. The more 'design space' we allow for the topology optimisation software, the better. If we start with an already minimal amount of material, the software does not have enough freedom to optimise the part.
2. **Apply a suitable material to the model**: Chose the material that you part will be made of. Remember that, by using topology optimisation, you are drastically reducing the amount of material used, so you may be able to use a more expensive and/or better material than the original material.
3. **Divide the model**: Split the model up into areas that you do not want the software to affect, and the 'design space'. The idea is to end up with the largest possible design space for the software to be able to optimise.
4. **Setup different scenarios** for your model of how forces might be applied. A good strategy can be to use a single force per scenario. Each scenario can be used to design an optimised part for that particular worst-case scenario. The various scenario design ideas can then be combined into a new design that covers all the force scenarios. If you understand the effect of each individual force, however, you can also setup the optimisation with multiple forces at the same time.
5. **Perform the topology optimisation**: Run the design through the software and let it do its work.
6. **Convert to smooth model**: Convert the rough TO results into a smooth printable model.

One of the most critical aspects of performing topology optimisation is that the forces that will act on the model must be well understood. If the model is not set up correctly, then the results it produces will not be meaningful (Fig. 3.23).

Note that topology optimisation is not a process that automatically produces ready-to-use models. It produces results in rough models that then need to be smoothed (Fig. 3.24).

There are various ways of producing such smooth models:

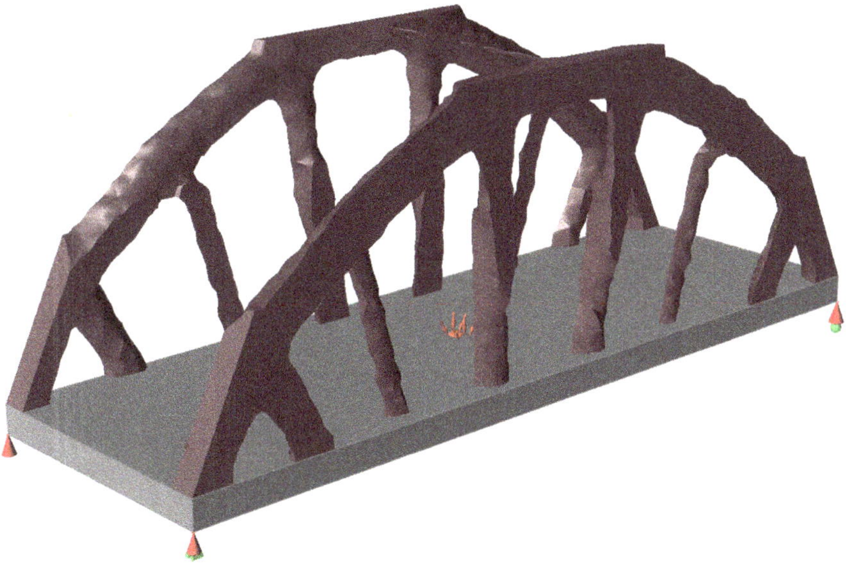

Fig. 3.23 A good understanding of the true forces and constraints acting on the part is essential for topology optimisation

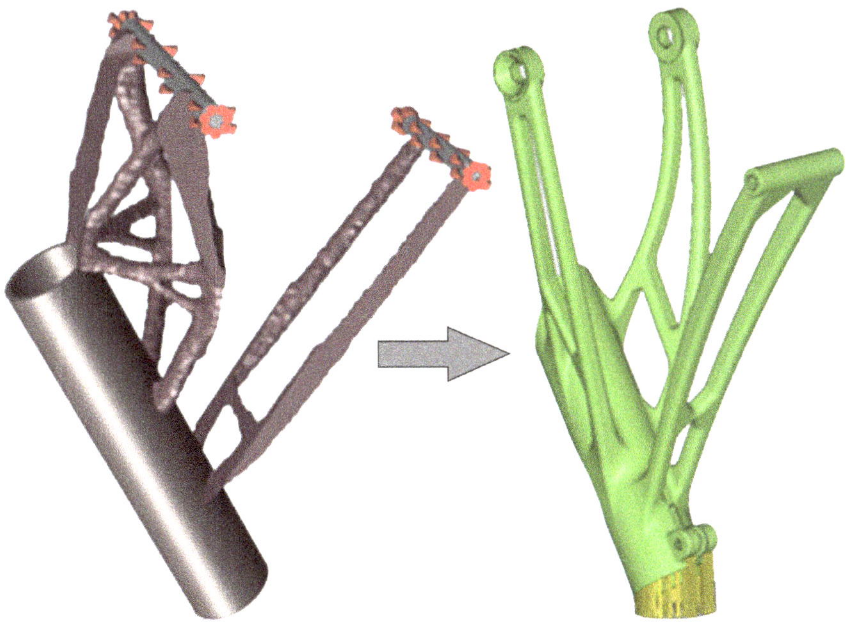

Fig. 3.24 Topology optimisation produces a 'rough' model that must then be smoothed, courtesy of Renishaw

- Use the built-in Polynurbs functions that some topology optimisation software is starting to offer. This allows you to smooth the models in a very organic way.
- Use specialized smoothing software such as Materialise 3 Matic or Geomagic.
- Use the rough topology optimisation design ideas as templates to Re-CAD the design in your native CAD. This is the preferred method of the authors of this book. The advantage of this method is that you end up with a parametric model that can be used in a conventional CAD manner.

Chapter 4
Computational Tools for Design Analysis and Optimisation of AM Parts

4.1 Aims of Using Design Analysis for AM

There are several possible purposes for using simulation tools related to AM. The first aim, just as with traditional design analysis, is to simulate the behaviour and performance of a virtual design, and to use this information to either manually or automatically improve the design according to some given criteria. This avoids the time-consuming and costly step of manufacturing a prototype and setting up test rigs, thus enabling rapid iteration in the design process. The second aim is to simulate the physical build process to aid in finding an optimal build orientation, support structures, material properties or to compensate for distortions.

4.2 Special Considerations for Analysis of AM Parts

The design analysis process is, in large, the same as for parts manufactured using more traditional methods. The same care needs to be taken to ensure that proper input data is given for the output to be meaningful. Most FEA software can be used for analysis of AM parts, however, there are some aspects worth mentioning that might often be overlooked when one is used to dealing with standard parts and materials.

4.2.1 Material Data

The material properties of an AM part are dependent on both the raw material, the machine type and the process parameters used for printing. The material supplier will give estimates of material properties such as Young's modulus, Poisson's ratio, yield strength, elongation at break, etc. in different orientations. However, the part

O. Diegel et al., *A Practical Guide to Design for Additive Manufacturing*, Springer Series in Advanced Manufacturing,
https://doi.org/10.1007/978-981-13-8281-9_4

geometry, process parameters and characteristics of each machine type will also affect the resulting material properties. For crucial applications, printing test specimens and characterizing the as-printed material properties might be necessary. Due to the inherent anisotropy of the as-printed material, it will often be necessary to include this in the material model and to specify the build orientation of the part.

Material properties are also dependent on the heat distribution in the part during the print, and, although post-processing heat treatment reduces these effects; residual stresses and deformations may have to be taken into account during the printing process. Software, such as Autodesk NetFabb, ANSYS exaSIM and MSC Software Simufact Additive can be used to predict and adjust for the distortion and residual stresses in AM parts.

4.2.2 Surface Finish

For applications where surface finish is of importance, such as where fatigue is a dimensioning factor, or for fluid analyses, care should be taken to properly characterize the surface properties of the as-printed part. As surface finish is highly dependent on orientation and location in the build chamber, fully describing this in the simulation tool may prove to be difficult. Although surface finish can often be improved through post-processing, hard-to-reach areas, such as internal voids, may be hard or impossible to machine or polish.

4.2.3 Geometry

Although so-called "complexity for free" is one of the major advantages of AM, complexity is not for free in terms of computational time when doing FEA. As with regular FEA, simplifying the geometry as much as possible is important to reduce the solution time.

4.2.4 Simplifying Geometry

Surface features such as text, logotypes, threads (in cases where simplified bolt joints can be used), non-structural rounds, etc. should be removed. However, even with superficial features removed, parts generated through topology optimisation and lattice-based designs can lead to part geometry with very high complexity, unfeasible for FEA due to the very large number of elements needed to represent the geometry. One approach to reducing this complexity is to approximate small-scale lattice structures with solid material with estimated material properties during the analysis or through beam elements. Reducing the complexity of topology optimisation based designs is often best done by manually re-modeling the part as a parametric CAD model and leaving out small-scale features.

4.2.5 Mesh-Based Versus Parametric Models

Software used for creating AM organic/high-complexity designs often only support a mesh-based workflow and geometry export. This can lead to software interoperability issues as FEA software require a volume mesh and not just a surface mesh (except for shells with a uniform wall thickness). Although the geometries meant for 3D printing should, ideally, be closed volumes, the requirements are not as strict as for FEA and the same geometry might print fine even though the analysis software may not accept it. Ideally, any part that needs to be analysed using FEA should be modelled/re-created in a solid modelling CAD software. However, it is not always feasible to manually re-CAD a complex model. Some software such as solidThinking Inspire and ANSYS DesignModeler allow for more or less automatic "skinning" of the mesh-based model to provide a polynurbs surface, as long as the input mesh is well-defined and without errors. For a single validating FEA this approach works well, however, should further design iterations be required, this process quickly becomes more time-consuming than re-modelling.

4.2.6 Geometry Distortion

For crucial applications, care needs to be taken to ensure that the analysed part geometry is the same as the printed geometry, i.e., any distortions due to the build process need to be taken into account, either through build process simulation or through accurate 3D scanning of the printed part. This is especially important for plastic parts with tight tolerances where contact pressures over time may lead to material creep and premature failure of the part.

4.3 Mesh

4.3.1 Parametric Models

In principle, no special consideration needs to be taken regarding the meshing process compared to regular FEA if the model is parametric. Ideally, the results should be mesh-independent, i.e., the results should not change if the mesh is further refined. Typically, the analysis software will have some means of conducting a mesh convergence study, where the mesh is iteratively refined in areas of high elementwise error until the results under consideration do not change appreciably between two refinement iterations.

4.3.2 *Mesh-Based Models*

For imported mesh-based models, care needs to be taken that the imported mesh is of sufficient level of detail for the FEA because information such as surface curvature is not typically available. Shell and beam elements should be used when possible to reduce the computational complexity of structures such as lattices and thin-walled designs.

4.4 Boundary Conditions

As with any FEA, accurate representation of the boundary conditions is required to get meaningful results. In structural analyses, point loads, fixed supports and prescribed displacements can lead to singularities with artificially high stresses. In fluid dynamic analyses, inlet and outlet boundary conditions typically poorly represents the flow in that region. The surfaces where boundary conditions are applied should thus be as far removed as possible from the areas of interest. If the geometry to be analysed is part of a larger assembly, consider including parts of the system to facilitate more accurate boundary conditions.

4.5 Optimisation

Several optimisation methods are applicable for AM designs, such as topology optimisation (TO), topography optimisation and size/parametric optimisation (PO). TO and PO have complementary strengths and weaknesses in terms of what can be optimised, the size of the design space and the amount of post-processing required to get a manufacturable design. The main application for TO is to reduce the mass of a part given some load cases. The result is in the form of a material field or mesh, most often used as inspiration during the concept development phase. PO, on the other hand, can be used to handle any computable constraint and/or objective and results in a parametric model with optimal dimensions. PO is typically used in the later stages of the development process to fine-tune the design.

4.6 Topology Optimisation

Topology optimisation is based on finding the ideal material distribution given certain objectives and load cases. Typically, this means finding minimum compliance structures, i.e., structures that, given a certain allowable weight, will be as stiff as possible. Although first introduced in the late 1980s, TO has largely been restricted to academic uses due to the high complexity of the generated designs and the computational cost of optimizing even moderately complex 3D structures. Even today, the organic results have restricted the use of topology optimisation to mainly be a means of inspiration to find optimal load paths in the concept design phase, even though the complex geometries can, in principle, be directly produced with AM.

Most TO algorithms are based on a discretized design space consisting of a number of two- or three-dimensional elements, with each element being assigned a pseudo-density. The pseudo-density represents the amount of material in a cell as number between 0 (no material) and 1 (fully dense material). The goal is to find an optimal pseudo-density for each element, given objectives and constraints. Intermediate material densities are typically filtered out through various penalization and averaging heuristics to avoid non-producible solutions.

4.6.1 Objective and Constraints

Topology optimisation allows for a number of different objectives, such as maximizing compliance (constraint on mass), minimizing mass (constraint on allowable stresses or displacements), maximizing resonant frequencies (constraint on mass), etc. Constraints can also be put on the geometry to facilitate different production methods, such as having a uniform thickness, draft angles etc.

The most common, and perhaps most robust, objective of TO is to minimize the mean compliance in the structure. The user specifies a certain mass, or volume percentage, that should remain in the design and the algorithm finds the optimal material distribution. The major disadvantage with this approach is that several iterations may be needed to find a suitable volume percentage.

4.6.2 Common Settings

A user controllable setting is often the minimum feature size, i.e., the thinnest allowable member in the optimised design. Without any restriction, the TO will tend towards a truss-like structure with many thin members, approaching Michell trusses (Fig. 4.1). While optimal, these structures are rarely suitable for manufacture, even with AM. Therefore, a minimum feature size should always be set to at least the minimum feature size of the AM machine.

The mesh-size, or detail level, in the TO will also greatly affect the outcome. Too coarse a mesh will lead to bulky designs, but too fine a mesh will require substantial time to solve and result in highly complex geometries.

4.6.3 Post-processing and Interpreting Results

Even with carefully selected values for the different settings, the output from a TO is most often not suitable for direct production. On a functional level, holes for fasteners and connections may have to be added back in. On an aesthetic level, the surfaces may have to be smoothened or made more symmetric. Although mesh-based editors can accomplish some of these tasks, a parametric modeler will streamline the process and

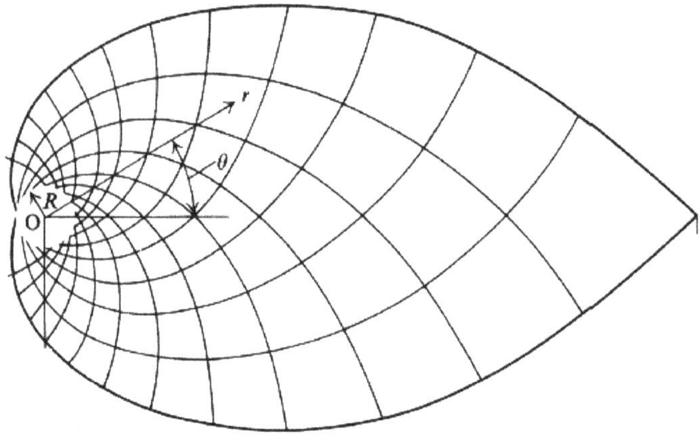

Fig. 4.1 Example of a Michell truss. Courtesy of Glaucio H. Paulino, Georgia Tech.

make the model future-proof by easily enabling modifications later on. Moreover, a parametric model will enable more straightforward design validation and PO.

Typically, the post-processing steps consist of filtering out cells with a material density below a user-configurable threshold. Another approach is to translate the continuous material density into mesoscale structures with variable density, such as lattices, that are possible to produce through AM. For instance, some software now allows for both TO and lattice size optimisation to further reduce the weight of the part.

4.7 Parametric or Size Optimisation

Topology optimisation is fairly limited in terms of what parameters can be taken into account during the optimisation. To fine-tune the design and to take other objectives into account, a parametric model and optimisation is needed. As this process is no different from PO of a part to be produced by traditional means, the reader is directed to any parametric modelling textbook for more information.

4.8 Build Process Simulation

As additive manufacturing, particularly in metal, is a time-consuming and expensive process, reducing the risk of build crashes and part failures is important. Although experienced operators and fine-tuned process parameters can alleviate some of this risk, there is a growing interest in build process simulation. Moreover, build process simulation allows for a better understanding of the localized material properties as a function of material, process parameters and part geometry. Some of the currently available commercial simulation tools that allow this include Autodesk Netfabb, ANSYS exaSIM and MSC Software Simufact Additive.

There are currently two main approaches to build process simulation: layer-by-layer simulation and scan pattern simulation.

4.8.1 Layer-by-Layer Simulation

Layer-by-layer simulation simulates the heating and cooling of each layer (several layers may be stacked together to reduce solve time) throughout the build process through a FE model. This simulation captures the residual stresses and deformation due to material shrinkage and can be used to predict whether the part distortion might interfere with the powder re-coater and the overall stress and deformation state of the finished part. It is also possible to simulate the effects of heat treatment and the removal of the part from the base plate. It is also possible to use the simulation data to automatically compensate the geometry for heat distortion, so that the finished part more closely matches the original CAD model. Additionally, the stress predictions can be used to automatically create support structures.

4.8.2 Scan Pattern Simulation

Scan based simulation also takes the layer-by-layer approach, but at a higher level of detail, enabling the laser scanning path to be taken into account. This provides more accurate results at the cost of simulation time.

4.8.3 Limitations

Although build process simulation offers many new possibilities, such as predicting part behaviour and compensating part geometry, the simulation is inherently sequential and is computationally expensive, resulting in simulation times that can currently sometimes match or exceed the physical build time.

Since the heating characteristics of the build is highly dependent on machine calibration, process parameters, material and part position in the build chamber (especially for polymer LS), it is crucial that these parameters are taken into account in the simulation. This calibration of the simulation model will have to be done by printing and measuring test artefacts using the parameters that are of interest for the simulation.

Chapter 5
Guidelines for Part Consolidation

Always design with the following thought in mind: With the function I am trying to achieve, what is the simplest possible configuration of part(s) that I can print in an orientation to avoid anisotropy? This thought often leads to good possibilities for part consolidation.

Part consolidation consists in transforming a product that is made up of many simple parts into one made up of fewer, but more complex parts. The guidelines below represent a general thought process that may help engineers and designers to decide whether part consolidation can be applied.

Think about the implications if you have a product made up of 10 components, and you can achieve a 10% part reduction? Maybe not much. But what if it's made up of 10,000 components? Or a million components? Now a 10% part reduction can have a significant impact. Now, what if you can achieve a bigger part reduction, such as 30% or even 60%? Some of the major impacts this can have include:

- Reduced engineering hours for assembly and installation drawings.
- Reduced material planning hours to load and get parts into purchase queue.
- Fewer parts to purchase (supply chain overhead).
- Fewer parts to receive/inspect (receiving O/H and quality hours).
- Reduced inventory burden (overhead).
- Reduced hours for pulling parts from inventory and kitting.
- Reduced manufacturing labour for assembly hours.
- Reduced inspection and certification steps.

The above improvements may offset the potential increased cost of AM consolidated components.

A very important aspect to remember in part consolidation is that there is no absolute correct answer. There are always many different possibilities that will allow a product to be made out of fewer components. But it is the context in which the product will be manufactured, assembled, and used, that determines whether one option is better or worse than another. It is the thought process one goes through with justifying whether a part should, or shouldn't, be combined with another that is important.

O. Diegel et al., *A Practical Guide to Design for Additive Manufacturing*, Springer Series in Advanced Manufacturing,
https://doi.org/10.1007/978-981-13-8281-9_5

5.1 Design for Function

First think of all the components that are there to perform a useful function in the product. Focus on the task that the product will perform in the particular context that the product will be used in. Optimise the design for its function rather than the processes used to make it. An example of this is shown in Fig. 5.1.

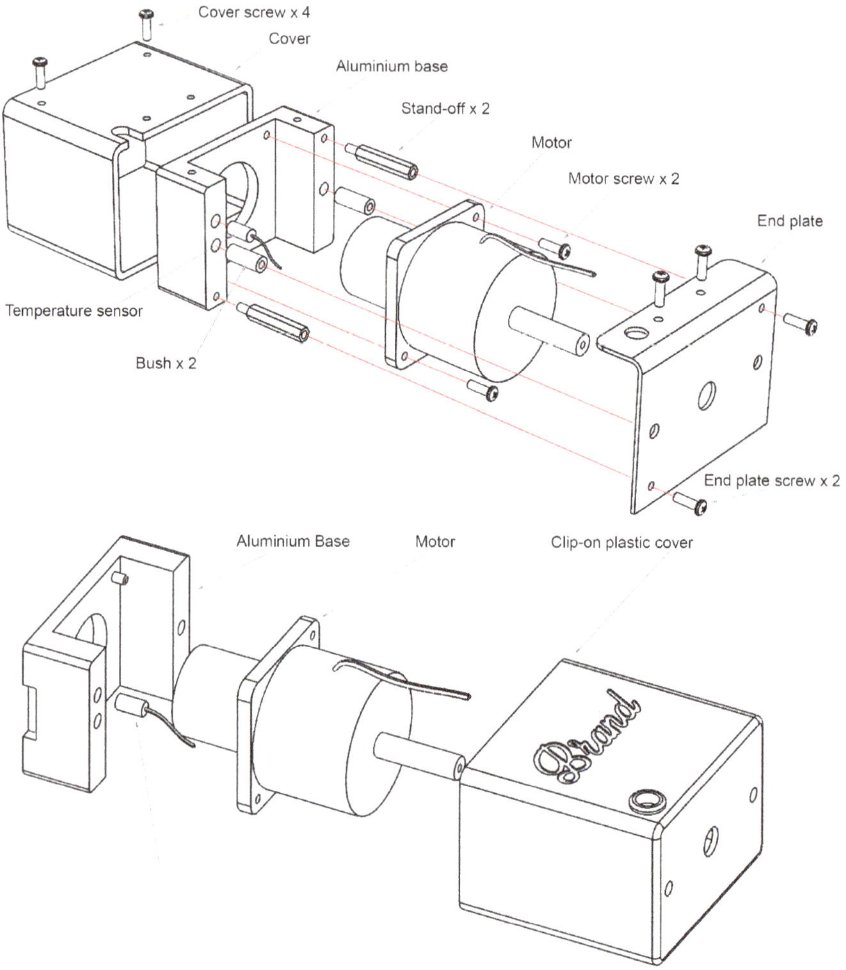

Fig. 5.1 The top design contains 19 components and several assembly steps and fasteners. The part-consolidated design reduces the part count to 4 components and allows for a much easier clip-together assembly

- Eliminate tight tolerance if at all possible. Holding tight tolerances can be costly but is often avoidable. If two mating parts are combined into one, all concerns, and costs, over controlling the tolerances where they mate vanish.
- Are all components that do not serve to add function superfluous?
- Can all components that are just there to mount or encase the components that perform a useful function be combined into a single component?

5.2 Material Considerations

If two or more parts are made of the same material, and do not move relative to each other, can they be consolidated into a single part?

If two parts that do not move relative to each other are made out of different materials, then:

- Are they made out of different materials just for historical reasons? If so, it may be worth considering changing them to be the same material.
- If they are different because of mechanical properties, can the part made out of the stronger (more expensive) material be made out of the weaker (cheaper) material and strengthened through ribs or hollow sections?
- If they are different because of thermal or chemical properties, can the part made out of the less resistant material be made out of the more resistant material?
- If the more expensive material has to be used, would fewer parts justify the material cost increase?

5.3 Number of Fasteners

If more than 1/3rd of the components are fasteners, then the number of parts and assembly logic should be questioned (Fig. 5.2).

Again, this example clearly demonstrates that part consolidation guidelines are not an absolutes that always guarantee a better product. In some contexts, possibly for assembly reasons, the assembly with many components shown on the left may be a better solution than that on the right. Also, because part volume plays an important role in determining the cost of AM parts, it may be more cost-effective to manufacture many smaller components and assemble them together despite the extra cost of labour and fasteners.

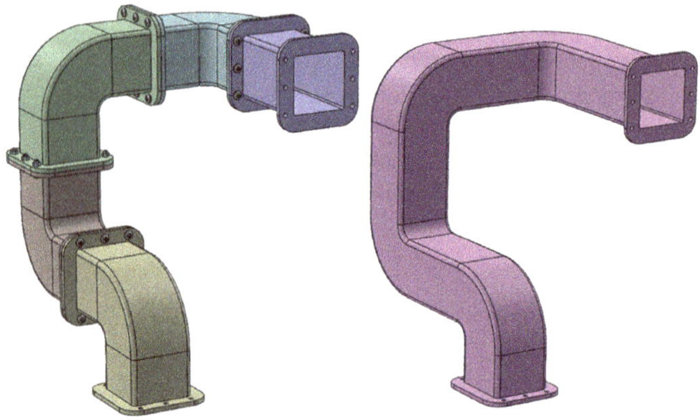

Fig. 5.2 The assembly on the left contains 48 fasteners (and potential twice as many if we use washers and lock washers) for 5 components. The design on the right eliminates all these fasteners, and avoids the risk of leaks between the seals

5.4 Use Knowledge from Conventional DFM/DFA

Most of the conventional design for manufacturing and assembly rules apply equally to additive manufacturing. AM just makes them easier to realize, and allows them to be taken even further than with conventional technologies (Fig. 5.3).

Again, in the above example, one cannot definitely ay that one design is better, or worse, than another. It all depends on the context in which they will be used. But it is always a good exercise to go through the mental thought process of considering different assembly and construction options and justifying the advantages and disadvantages of each.

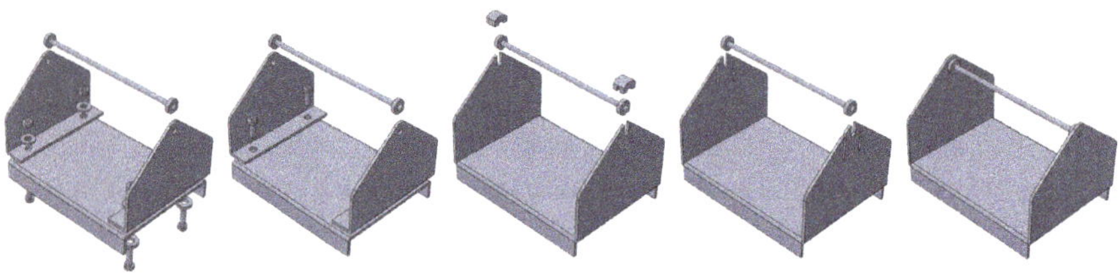

Fig. 5.3 Four different DFM/DFA strategies for conventional manufacturing and one for AM

5.5 Assembly Considerations

A danger of part consolidation is that designers go 'over the top' and design heavily part consolidated products, but that are difficult and time-consuming to assemble.

- With conventional manufacturing, parts are sometimes subdivided to make assembly easier.
- With AM consolidated parts, even greater care must be taken so the product can be assembled.
- To aid assembly, could areas of flexibility be introduced into a complex consolidated part to allow for easy assembled?

5.6 Moving Parts

AM makes it possible to produce assemblies with moving parts (mainly in plastic). But consideration should be given as to whether it is practical for real engineering applications.

AM processes have a relatively low level of precision and require relatively large gaps between moving parts. It is therefore impossible to make a working ball-bearing (other than as a decorative object), for example, but it is possible to make a working tie-down that requires a relatively low level of precision (Fig. 5.4).

- The clearance between moving parts is very dependent on the area of the surfaces that will be in close contact. This is discussed further in the design for polymer AM section of this book.
- A small surface area is more forgiving than a large surface area, and requires a smaller gap between the moving parts.

With additive manufacturing, there is often a temptation to make products with moving parts just for the sake of having a product printed with moving parts, rather than because the moving parts add some value. Do not create moving assemblies only for the sake of having them.

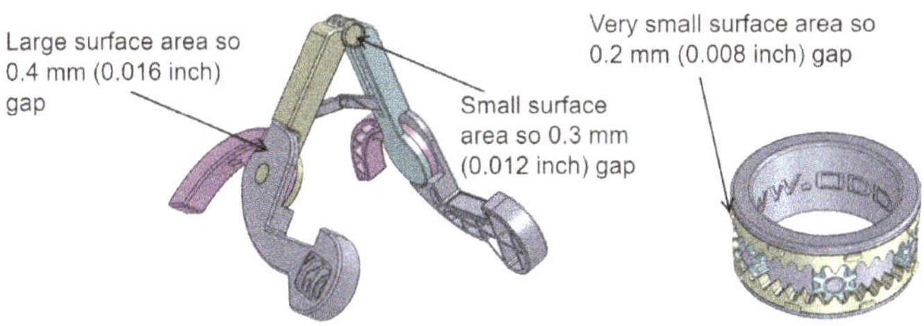

Fig. 5.4 Required gaps between moving parts is dependent on surface area

5.7 Common Sense

As with all engineering, common sense is important. Consolidate parts only when there is a benefit to doing so. And, because it is so important, we repeat the point made above about part consolidation being a thought process rather than an absolute set of design rules.

Part consolidation can also be a good exercise in justifying materials used. Just the process in questioning why particular materials are being used can often lead to a reduction in the number of materials used in a product.

Chapter 6
Guidelines for AM Tooling Design

We have, so far, mostly been talking about using AM for direct part production. However, many industries are now starting to use AM as a way of manufacturing the tooling for conventionally injection molded, cast, extruded, or sheet-metal parts. This means that the final parts are exactly the same as those that they would have previously made with conventional tooling, and only that the tooling, itself, was produced with AM. This makes it relatively easy for engineers to accept the technology, as the produced parts are identical (or better) to the parts they are used to.

In many cases, AM can produces tools in a matter of a few weeks, compared to several months with conventional manufacturing. Because of the so-called 'complexity-for-free' advantages of AM, tooling can be made to produce parts more efficiently and of better quality than conventional tooling. This chapter describes some of the design guidelines that can help in producing more efficient AM tools.

6.1 Mounting Fixtures and Guides

Because of the relatively rough surface finish of metal AM parts, in most cases, your tool will need a quick finishing cut on a CNC machine to give it the desired surface finish. In order to make this process as easy as possible, a few simple guidelines can greatly help:

- Allow excess material in critical areas of your AM design so this can be machined off. Usually, adding 0.5 mm extra material is enough.
- Build in mounting fixtures into your design so it can be quickly and easily mounted in the CNC machine for the finishing operation. This is important, as the part of the CNC machining operation that may require the most man-hours is setting up the AM part in the machine, clocking it straight, and accurately setting its origin.
- Even the addition of simple slot fixtures mean that your tool will, at least, be inserted into the CNC machine straight and aligned in 2 axes in the matter of only a few minutes (Fig. 6.1).

O. Diegel et al., *A Practical Guide to Design for Additive Manufacturing*, Springer Series in Advanced Manufacturing,
https://doi.org/10.1007/978-981-13-8281-9_6

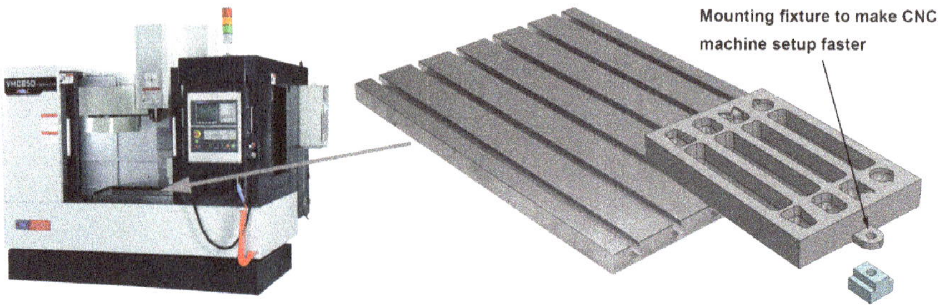

Fig. 6.1 The simple addition of mounting/fixture points to a part can make it a relatively easy operation to setup and clock in a CNC machine

6.2 Conformal Cooling

Part cooling is an important part of the process to produce quality injection molded parts but can sometimes consume from 50 to 80% of the cycle time per build. By adding conformal cooling channels to your injection molding tool, you can greatly increase its running efficiency, improve part quality, and extend the tool life.

Conformal cooling channels are curved channels that allow coolant to more efficiently get to the parts of the tool that need to be cooled. With conventional tooling, cooling channels are made by drilling straight holes, some of which then get plugged, to produce a cooling channel through which coolant can flow. With additive manufacturing, complex curved channels that closely follow the contour of the part to be molded, referred to as conformal cooling channels, can be made to get cooling precisely where it is needed (Fig. 6.2).

Conformal cooling channels, applied with no engineering simulation or analysis will, generally, result in about a 10% cycle time improvement. Conformal cooling channels, applied with engineering simulation and analysis will, generally, result in cycle time improvements from 20 to 40%. That considerable cycle time reduction can equate to substantial increased profit.

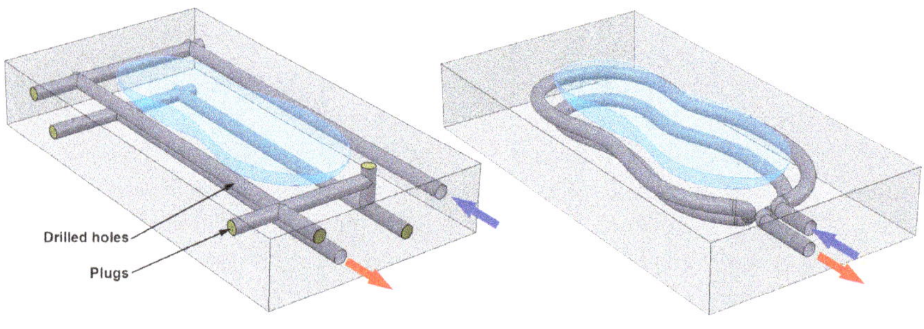

Fig. 6.2 Conformal cooling allows for much more complex cooling channels than is possible with conventional drilling techniques

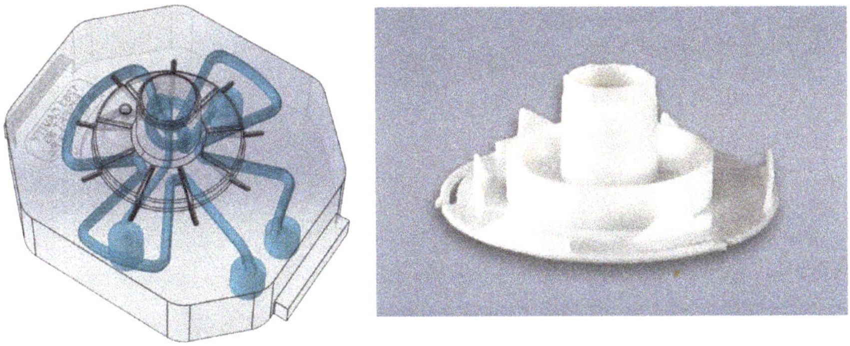

Fig. 6.3 Plastic injection molded part and conformally cooled mold, courtesy Phillips Plastics (now Phillips Medisize)

In the example below, conformal cooling is used to improve cycle times. The part is an injection molded component made by Phillips Plastics. With a conventionally made injection molding tool, the cycle time with a single cavity mold is 16.78 s. In contrast, the cycle time with a 4 cavity AM tool with conformal cooling is 13.02 s. This represents as cavity cycle time improvement of 22.4% (Fig. 6.3).

When plastic cools evenly, internal stress is minimized. This results in a higher quality part with less warping or sink marks. The more controlled cooling offered by conformal cooling channels allows you to precisely control how the plastic solidifies in the mold and, therefore, to minimize part distortion and shrinkage (Fig. 6.4).

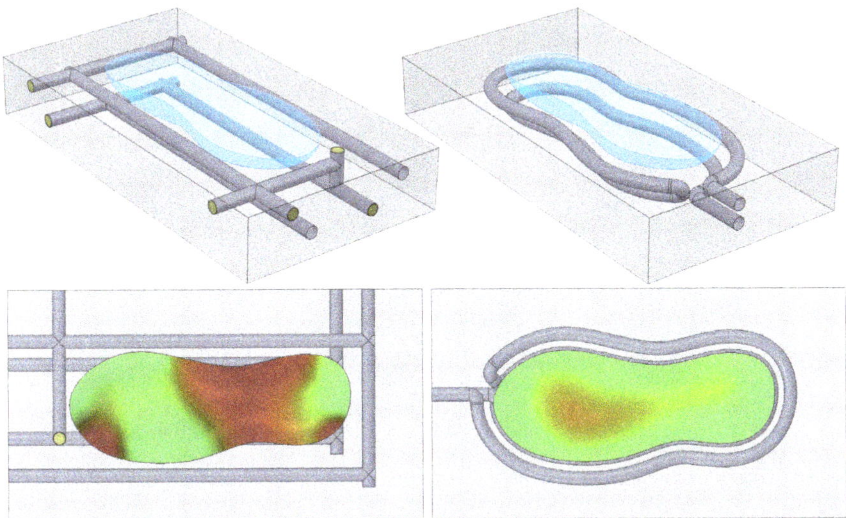

Fig. 6.4 Heat distribution comparison between conventional and conformal cooling channels

In the previous Philips Plastics injection molded part example, the conformal cooling tool produces better parts.

The flatness specification used for the part made with conventional tooling is to have a deviation of less than 0.25 mm. The produced parts have a flatness deviation of between 0.15 and 0.223 mm.

In contrast, the flatness specification for parts produced with the AM tool with conformal cooling was reduced to being better than 0.2 mm, and the actual measured flatness deviations were measured to between 0.080 and 0.161 mm. This represents a quantifiable part feature improvement of 20%

6.3 Coolant Flow Strategies

When designing conformal cooling channels, the first decision that needs to be made is which coolant flow strategy to use (Fig. 6.5):

- A zigzag pattern, also known as a series cooling path, has part regions cooled one after the other rather than at the same time. Cooling in series is generally not preferred unless parts are small enough that the delay is negligible.
- A parallel channel design allows for different areas of the mold to be cooled at the same time. The main drawback of the parallel cooling method is that it requires a lot of coolant.
- A spiral conformal cooling channel design is often used with parts that have curved or spherical elements.

On complex tools, one can sometimes combine cooling strategies where, for example, part of the tool uses a zigzag type strategy, while the rest of the tool employs a parallel strategy.

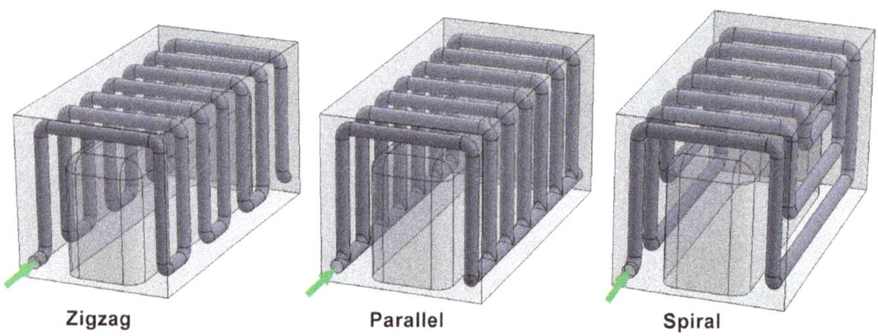

Fig. 6.5 Types of cooling strategies that can be employed with conformal cooling

6.4 Coolant Channel Shape

The coolant channel shape can affect both the ability of the AM system to effectively produce them, as well as their effect on cooling efficiency.

From experience, the optimal diameter is usually somewhere between 4 and 12 mm (depending on the design of the product). The diameter should, however, be carefully chosen depending on the AM system being used. Remember also that round horizontal channels, for example, will require internal support material if their diameter is above 8 mm diameter. The most common channel shape is, generally, round but, on occasion, vertical elliptical holes, or house-shaped or teardrop-shaped channels are also used.

The cooling performance can, sometimes, also be increased by ribbing the shape of the channel which causes an increase in the expected turbulence in the channel (higher Reynolds number), which thus increases cooling (Fig. 6.6).

From the authors personal experience, the vast majority of cooling channels are round, and usually between 4 and 8 mm in diameter. This could, in large part, be due to the fact that round channels are the simplest to CAD, particularly for parallel type strategies, where round joint intersections are much easier to control than non-round intersections.

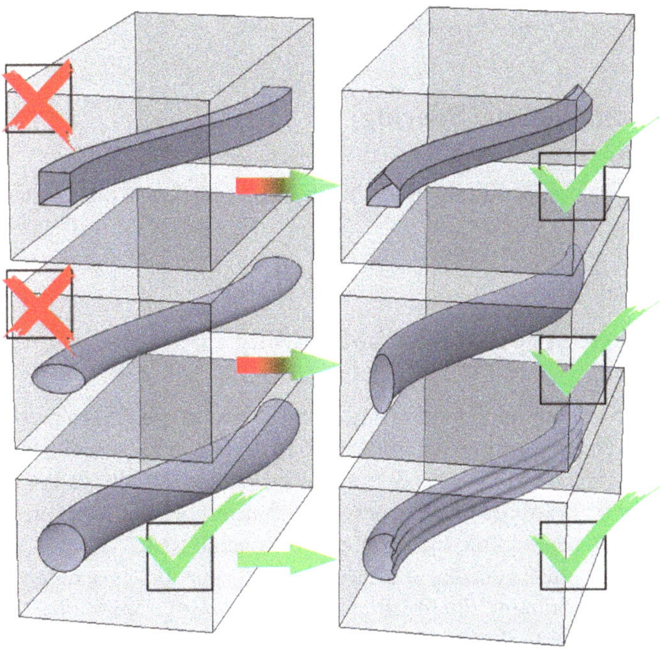

Fig. 6.6 Channel shapes can affect ease of build, as well as cooling effectiveness

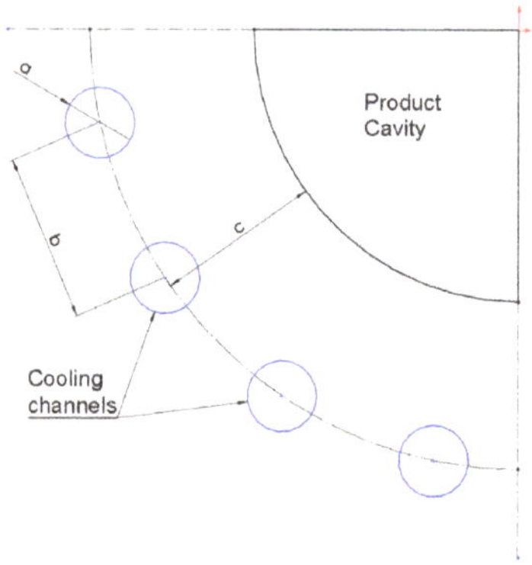

Wall thickness of molded product (mm)	Hole diameter a (mm)	Centre distance between channels b	Distance from channel centre to cavity edge c
0 – 2	4 - 8	2 to 3 * diameter	1.5 to 2 * diameter
2 – 4	8 – 12	2 to 3 * diameter	1.5 to 2 * diameter
4 - 6	12 - 14	2 to 3 * diameter	1.5 to 2 * diameter

Fig. 6.7 Guidelines for conformal cooling channel spacing

6.5 Coolant Channel Spacing

For coolant channel spacing, the following guidelines make for a good starting point (Fig. 6.7).

6.6 A Hybrid Approach to AM Tooling

Like with most conventional tooling designs, it is best to use AM to produce only a tool insert that drops into a conventional mold base. This is because the mold base is, essentially, a large block of steel and, from the chapter on the economics of AM, you will remember that it would be extremely time-consuming, and therefore expensive, to produce, while adding little to no value to the overall tool.

You may also often find that if a large part of your tool is a relatively simple geometry that can be more efficiently manufactured with a traditional technology. In that case use a hybrid strategy. A hybrid strategy is one in which part of the component is made through conventional manufacturing, and then used as the build plate, so the AM system builds the AM part of the component directly on top of the conventionally machined component.

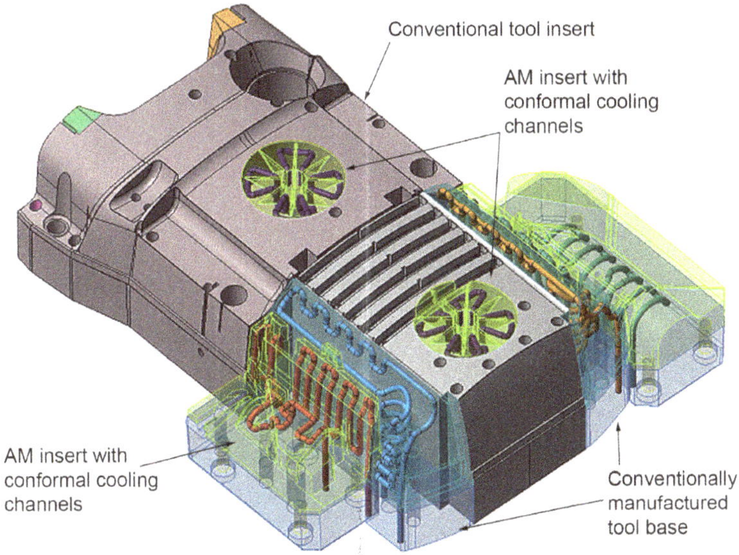

Fig. 6.8 A hybrid approach to tooling where the complex part is grown directly onto the machined simple part, courtesy of Renishaw

In the case of tooling, for example, it may be worth splitting your tool into its complex component, that is best manufactured through AM, and its simple component that is best manufactured by conventional methods. The simple part can be mounted into the AM system, and the complex part of the tool can be built directly onto the simple part (Fig. 6.8).

6.7 Minimise Print Time in Tooling

As discussed in the economics of AM chapter, large masses of material are slow and expensive to produce. Tooling is a typical application in which the bulk of the tool is a large mass of metal which servers little purpose. It is there because, with CNC machining we try to minimize the amount of cutting that needs to happen. With AM we have the opportunity of creating much lighter tools that have an even wall thickness of metal, and take less time and cost to manufacture.

Using the tool for a shoe insert used in the conformal cooling section above as an example, we can see that the vast bulk of the tool is a solid mass of steel. Could this, instead, be redesigned as follows as a way of reducing the amount of material used? Whether this can be done, will of course depend on many factors such as the pressure the tool will be subjected to, etc. But, in many applications, a wall thickness of 10–20 mm is more than adequate, and still leaves enough material for conformal cooling channels (Fig. 6.9).

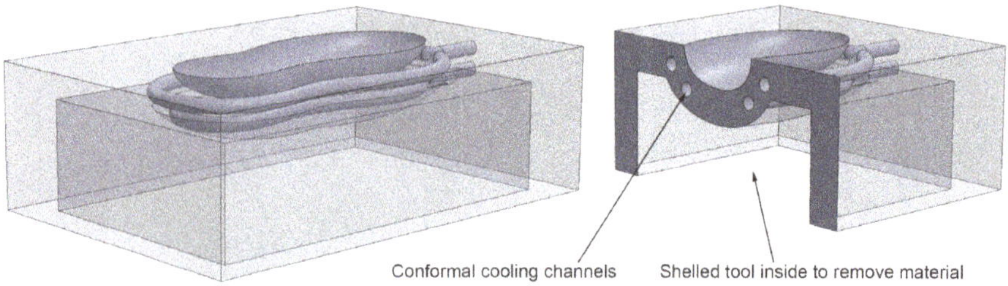

Fig. 6.9 Example of improving the print time and cost of a tool by shelling its interior

Fig. 6.10 Honeycomb structure used on indie of sheet metal tool to reduce its print time and cost, courtesy of 3D MetPrint AB

Other options to achieve a similar goal would including filling the inside of the tool with a honeycomb structure, or a lattice (Fig. 6.10).

Chapter 7
Design for Polymer AM

The design guidelines in this section apply to almost all polymer AM technologies. Some technologies have specific guidelines that apply only to that technology, and these are discussed in the chapter on design guidelines for specific AM processes.

7.1 Anisotropy

Anisotropy is the term used to describe the properties of a part in which the mechanical properties of the part are not the same in all directions. With all additive manufacturing technologies, there is always a certain amount of anisotropy in the vertical direction, between the layers. This is because the mechanical strength of the bond between each layer can be somewhat weaker than the mechanical strength within the layer itself. In some AM technologies, this anisotropy can be negligible or eliminated through post-processing, but in others it is an issue that must be take into account when designing a part, and when deciding on its print orientation.

As all AM technologies suffer from some level of anisotropy, special consideration needs to be given to all aspects of the design that could suffer from weakness because of print orientation. A clip, such as that shown below, for example, should, if possible, never be printed in the vertical orientation as it is likely to break when in use.

O. Diegel et al., *A Practical Guide to Design for Additive Manufacturing*, Springer Series in Advanced Manufacturing,
https://doi.org/10.1007/978-981-13-8281-9_7

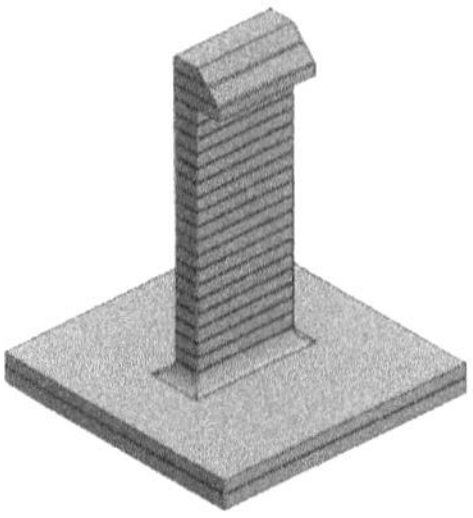	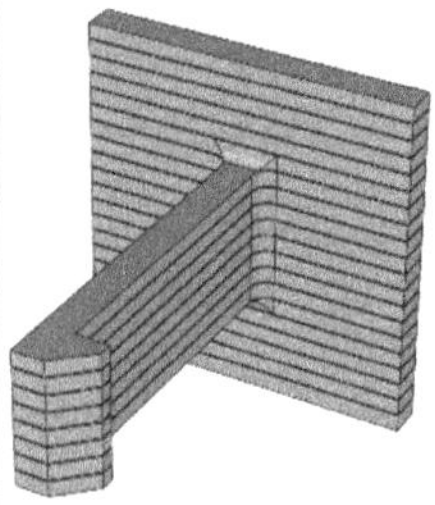	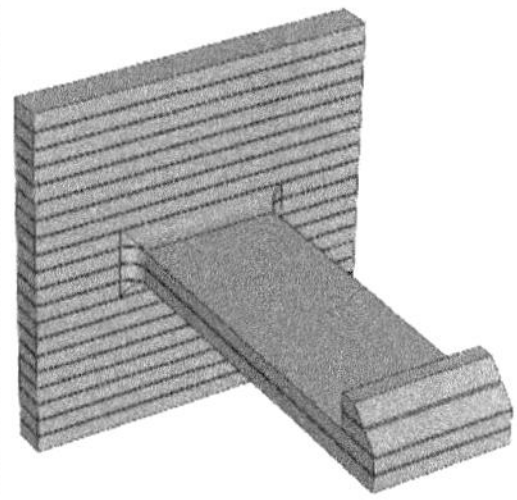
Clip will be weak and, almost certainly, break	Good compromise clip, with decent spring and strong hook	Clip has the best spring strength and flexibility but a weak hook

Some technologies, such as powder bed fusion, have some degree of anisotropy but, as you get past a certain thickness, the anisotropy gets minimized. This is due to the larger mass of plastic retaining the heat for longer, which creates a better bond with the layer below. With powder bed fusion, screw bosses, below a diameter of about 6 mm, for example, still exhibit some anisotropy, whereas it become negligible for screw bosses with a diameter greater than 6 mm.

As geometries becomes more complex, however, print orientation often becomes a compromise between avoiding anisotropy while, at the same time, achieving the best surface finish, and the best mechanical properties for as many as features as possible.

7.2 Wall Thicknesses

In general, wall thicknesses with polymer AM parts should be similar to those used if the part were injection molded. For light-weight consumer products, this ranges from around 0.6–2.5 mm, and for more industrial heavy-duty industrial products, this can range from 3 to 5 mm. Though it is possible to create thinner walls, how successfully they will print will depend on the surface area of the wall, and the unsupported width to height ratio.

Large surface area flat thin walls will be hard to print without distortion and, depending on the AM technology used, may delaminate. A simple technique to avoid this problem, if the wall cannot be made thicker, is using ribs to reinforce the wall.

As with conventional injection molding, a general rule of thumb is to use even wall thicknesses throughout the parts, as uneven wall thicknesses can create part distortion. The design freedom allowed by AM, especially in comparison with injection molding, makes it much easier to achieve such even wall thicknesses throughout the part. This does not mean that there cannot be uneven wall thicknesses, only that there should be a really good engineering or functional reason for doing this.

Fig. 7.1 Printing a large flat surface at an angle reduces the surface area being melted in any one slice and, thus, reduces the risk of distortion

In some case, the orientation in which the part is printed can be used to prevent large flat walls from warping. In the simple box example below, if the part is printed in the horizontal position, then there will be a large thin 'sheet' of polymer that gets melted, and may try to curl up and cause warping, or could even cause the machine to crash. In contrast, printing the part at a slight angle, typically above 10°, removes such large flat areas and can drastically reduce the risk of distortion (Fig. 7.1).

But, as with all AM guidelines, this too is a compromise, as printing the part at an angle may result in poorer surface finish than if it is printed straight.

For recommended wall thicknesses for each particular AM technology, refer to the section of this book on design guidelines for specific AM processes.

7.3 Overhangs and Support Material

With polymer AM technologies, almost all technologies, with the exception of powder bed fusion, and some binder jetting technologies, the printed parts require support material to support any overhanging features. Support material is a sacrificial material that is utilized during the printing process to allow any features that overhang, because it is not possible to print in air without the material collapsing, and is removed after the part has finished printing (Fig. 7.2).

Most AM systems allow you to choose at what angle to use support material. It takes some trial and error to determine what angle is best to use to get the best quality part while, at the same time, minimizing the amount of support material used.

There is usually a support 'angle' option in your 3D printing software that determines the angle in which the part requires support material. Some printers measure this angle from the vertical, while others measure it from the horizontal. It is therefore important to be aware of how each particular printer takes this measurement (Fig. 7.3).

Choose the orientation that your part is printed into optimise support material use but, in doing so, keep anisotropy in mind, as one print orientation may give you less support material, but may weaken the part in certain undesirable areas.

With the part below, for example, if it is printed facing upwards, very little support material is used, whereas if it is printed upside down, the inside will be filled with

Fig. 7.2 Support material to allow overhanging features to be printed

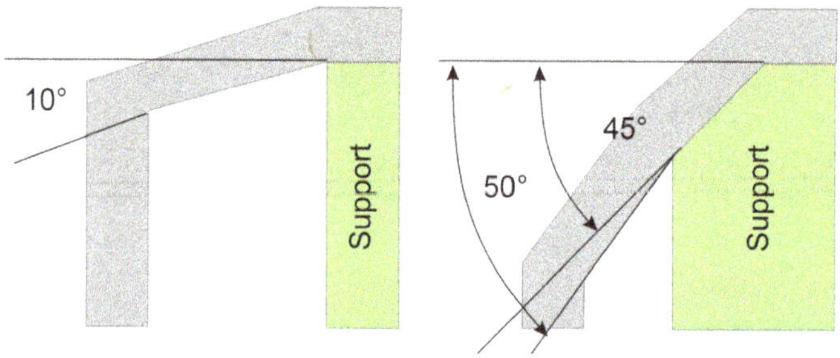

Fig. 7.3 Setting the angle at which to use support material

material, which requires labour to remove after the part is printed, and wastes material. Remember, also, that it is always easier to remove material from the outside of a part than from the inside. The surfaces where the support material makes contact will also, generally, have a worse surface finish than the upward facing surface, so will require extra post-processing to make smooth (Fig. 7.4).

Some AM software also allows you to set the surface area below which no support material is required. The benefit of this is that some material is saved and a slightly shorter print time is achieved. Moreover, this approach can also result in less support material to remove from holes. The risk is that, if the surface area is too high, there may be some sagging of material on the top surfaces of the overhangs (Fig. 7.5).

7.4 Holes

With AM technologies, print orientation can greatly affect the roundness of holes. To achieve the roundest possible holes, it is always best to have the holes printed in the

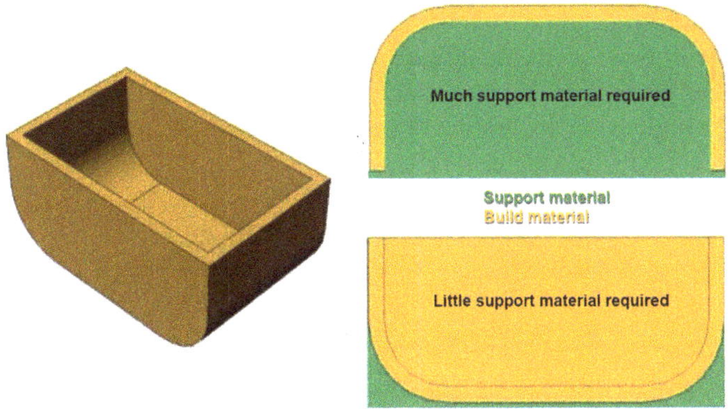

Fig. 7.4 Support material versus print orientation

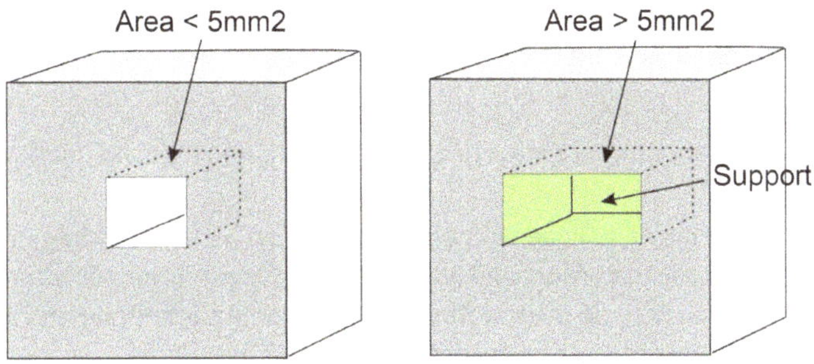

Fig. 7.5 Setting the surface area below which support material is not required

vertical direction. Holes printed in the horizontal position will suffer both from the stair-step effect, and from some sagging that may make the holes slightly elliptical.

Holes are also often very slightly undersized, but this can easily be compensated for in CAD by oversizing the holes by about 0.1 mm (but test this for each particular printer as every printer will give slightly different results), or by running a drill through the hole after printing to make it the precise required size.

The smallest achievable hole diameter depends heavily on the thickness of the material they are going through. See the picture in the #1 rule of design for AM section for a visual explanation of this phenomenon. In general, however, for most walls that are around 2 mm thick, a 0.5 mm diameter hole is achievable.

7.5 Ribs

Most polymer AM materials are slightly less rigid than their injection molded counterparts. This means that large surface areas and walls can be quite flexible, and can sometimes develop some distortion during the printing or cooling down process. The simplest way to make walls more rigid, and to minimize the risk of distortion is to design the part with ribs to reinforce large thin areas.

The general guideline for adding ribs to a 3D printed polymer part are:

Thickness of ribs: 75% of wall thickness	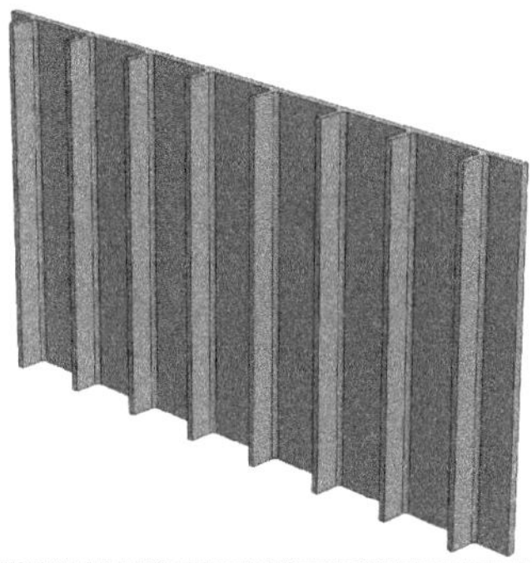
Height of ribs: <3× thickness	
Rib spacing: >2× thickness	
Always fillet the point where ribs meet the wall	

For a given stiffness, it is better to increase the number of ribs instead of their height.

For very thick ribs, it is better to core them out, so as to avoid large masses of material that can cause distortion, and are more expensive to print. Other options can include shelling the rib to an even wall thickness (making it hollow) and printing it filled with support material, or with a lattice structure (Fig. 7.6).

7.6 Avoiding Superfluous Material

With conventional subtractive manufacturing, we strive to have the machine do as little cutting work as possible so, in our design, we leave any material that does not hinder the function of the part, as removing it would cost time and money. With AM, however, it's the other way around. The more unnecessary material there is, the more work the AM system has to do, and the longer the part will take to make, and the more it will cost. Refer to the chapter on the economics of AM for more details on this.

Large masses of superfluous material can also have a detrimental effect on the part, as they may cause the part to distort and warp on cooling.

When designing AM parts, it is therefore important to avoid having large masses of material that serve no functional purpose, as they slow down production time, increase part weight, and can cause part deformation. The very simplest approach to this, also described in the economics of AM section, is to 'shell' the thicker sections of the part. This will minimise print time and cost. But a decision also needs to be made as to whether to leave the excess material (unsintered powder, liquid photopolymer resin, or support material, etc.) inside the shelled part, or to design in

Fig. 7.6 Coring out thick ribs

'salt-shaker' holes so that the excess material can be removed. To remove internal support material, larger holes may be necessary (Fig. 7.7).

Fig. 7.7 Shelled part with salt-shaker holes to remove thick sections from the part and allow material inside the part to be removed

7.7 Font Sizes and Small Details

7.7.1 Small Details

The minimum size for small detail to still be visible is determined by the printer's resolution. When detail dimensions are below the minimum, the printer may not be able to accurately replicate them.

Details that are too small can also be smoothed over in the polishing or sanding process. To ensure details come out clearly, make them larger than the indicated minimum.

Typically, for most polymer AM technologies, details are visible down to about 0.5 mm (though in some cases they can be as small as 0.2 mm high × 0.2 mm wide), but this must be tested for each model of printer. Also, surfaces that are in contact with support material may not be possible to reproduce with as fine a detail as those surfaces that do not require support material.

7.7.2 Font Sizes

For many AM technologies, the smallest legible font size is, counterintuitively, on the sides of the part. Relatively small text can be added to the vertical sides of a part, but it can be relatively poor on top surfaces.

Fonts, and other small details, can either be sunk into the wall (debossed) of the part or can protrude from the wall of the part (embossed). In general, it may be preferable to have them sunk into the walls of the part for two reasons: Firstly, it removes material form the part which means a slightly reduced print time and, second, it reduces the risk of the font or details being sanded off during post-processing. However, there is no problem with using embossed fonts, if required. But greater care may need to be taken during the part post-processing.

A font size that usually works on all surfaces is 14pt, and at least 0.4 mm (0.016 in.) in depth. On vertical surfaces one can go down to about an 8pt font (Fig. 7.8).

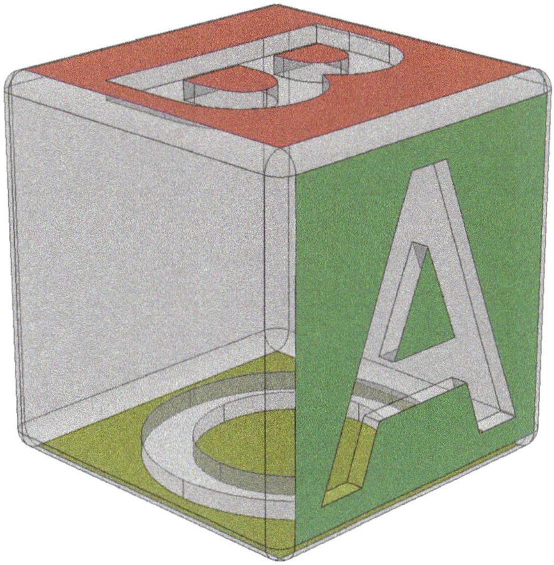

Fig. 7.8 Surfaces with best quality font sizes for polymer AM

Chapter 8
Polymer Design Guidelines

8.1 Designing for Material Extrusion

Material extrusion (a.k.a. fused deposition modelling or FDM) is an AM process in which material is selectively dispensed through a nozzle much like a hot-glue gun. Materials used are typically thermoplastic polymers and the part being constructed normally requires support structures for overhanging features. Some system use support material built from the same material as the print material, some use a different second material that is slightly easier to remove by mechanical force, and some use a second material that is soluble.

8.1.1 Material Extrusion Accuracy and Tolerances

There is a vast difference in accuracy and tolerance between different material extrusion systems. They also vary depending on geometric features and print orientation. The only sure where to know the accuracy and tolerance of any particular system is to print a test reference part and to measure it.

- Accuracy is how close the part is to the CAD model data.
- Tolerance is the acceptable degree of variation.

The numbers given below are for industrial quality material extrusion systems. They represent a general tolerance and accuracy for material extrusion technologies.

Layer thickness	0.1–0.3 mm (0.005–0.013 in.)
Accuracy	±0.1 or ±0.03 mm per 25 mm (±0.005 in. or ±0.0015 in. per inch), whichever is greater
Tolerance	Reality rule of thumb for Material Extrusion: typically 0.25 mm (0.01 in.)
Smallest feature size	Around 1 mm (0.04 in.)

O. Diegel et al., *A Practical Guide to Design for Additive Manufacturing*, Springer Series in Advanced Manufacturing,
https://doi.org/10.1007/978-981-13-8281-9_8

8.1.2 Layer Thickness

The first decision that has to be made when printing material extrusion parts is what layer thickness to use. In general, the thinner the layer, the better the surface quality, particularly on rounded parts, as the stair-step effect will be much less visible. However, the thinner the layer, the longer the part will take to print. A 0.1 mm layer thickness will take three times longer to print than a 0.3 mm layer thickness.

If a part is mainly composed of flat geometric features in the vertical direction, then printing it with a thicker layer thickness will not produce much worse surface finishes that a thin layer thickness but will print much faster. If the part is made up of many curved surfaces, then a thin layer thickness may be preferable in order to achieve curved surfaces that are as smooth as possible.

8.1.3 Support Material

Another decision that has to be made when printing material extrusion parts is what type of support material options to use. Almost every different maker of material extrusion printer offers different options for this. But some of the common ones include (Fig. 8.1).

In addition, some systems also allow some smarter choices of support structures, such as the one shown in Fig. 8.2.

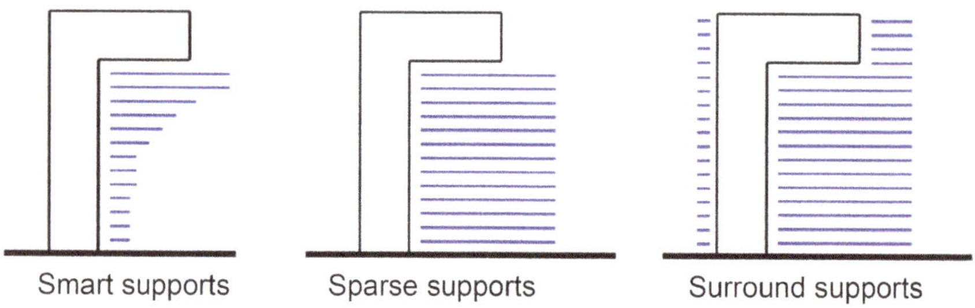

Fig. 8.1 Types of support material

Fig. 8.2 Examples of different support material densities

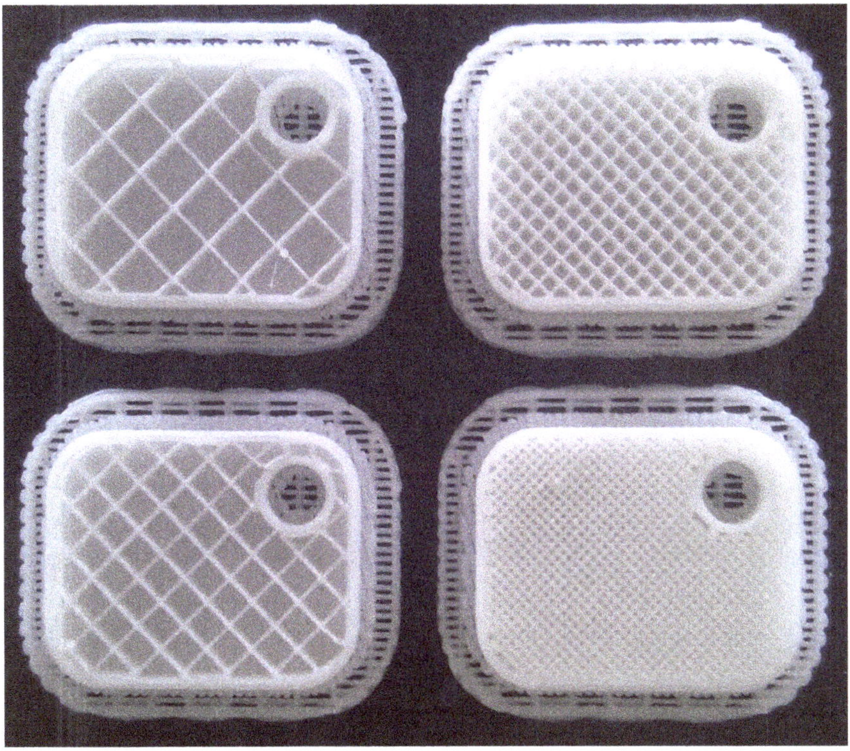

Fig. 8.3 Examples of different interior fill options

If the supports are removed manually, rather than being dissolved away, the designer should allow access to the supports. Also, care should be taken when creating small features that will be attached to the supports because they can break accidentally when the support is being removed.

8.1.4 Fill Style

Most material extrusion allow the user to decide whether the part should be printed as a solid part, or as a 'sparse' part in which the interior void is filled with a scaffold structure. Some systems also allow the user to specify how thick the outer shell wall should be. The systems usually also allow the user to select the infill percentage, that is, how dense the scaffold structure should be. An infill percentage of 0 means there is a void (so this is a way of letting the 3D printer software automatically shell the part), and 100% means completely solid material. In terms of mechanical strength, infill percentages above 50% can have diminishing returns (Fig. 8.3).

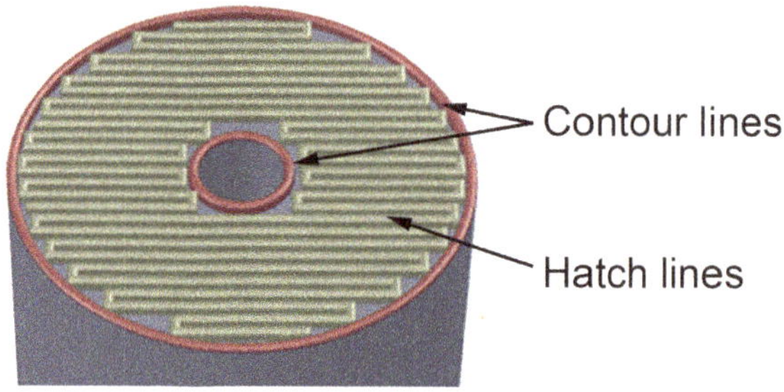

Fig. 8.4 Reinforcing the joint between contour and hatch lines

Parts created using polymer material extrusion usually have anisotropic material properties, i.e., the part will be weaker in the Z direction than it is in the X and Y directions. Designers must take this into account by either designing to the lower material properties specification, or by ensuring that highly stressed features are built in a horizontal rather than a vertical direction. This requires close cooperation with the AM process operator.

8.1.5 *Other Considerations*

Another characteristic of this process is a pronounced "stair-stepping" on the surface of parts. This can be reduced by various post-processing techniques (such as using an acetone vapour chamber for parts printed in ABS), but this will have an impact on part accuracy and sometimes material properties. See the post-processing chapter for details on a variety of finishing options.

Material extrusion AM can, on some certain wall thicknesses, leave a small air gap between the laid down filament. This is caused by the software having to make a decision on whether or not to deposit an extra strand of material in the wall. If, for example, the strand of polymer exiting the printer nozzle is 0.4 mm wide, but the wall thickness is 0.9 mm, the software must make a decision as to whether or not to push an extra strand of polymer between the two 0.4 mm tracks. This is machine brand and model dependent, so the recommendation is to do some tests and to find the few wall thicknesses that are not desirable for that particular model of machine.

Holes in a Material Extrusion part are generally undersized. For tight tolerances, drill/ream the holes to the exact diameter, or compensate in CAD by 0.2 mm (0.008 in.) diameter.

Because of the somewhat weak bond between contour lines and infill (or hatch) lines, self-tapping screws can sometimes strip away the contour material inside of screw bosses. A drop of super glue, allowed to wick between the contour and fill material can help to resolve this (Fig. 8.4).

The following pages contain information on how to design specific features for the material extrusion process.

8.1.6 Feature Type: Vertical Wall Thickness

Process variable	Wall thickness (t)	
Layer thickness	Minimum	Recommended minimum
0.18 mm (0.0071 in.)	0.36 mm (0.014 in.)	0.72 mm (0.028 in.)
0.25 mm (0.0098 in.)	**0.50 mm (0.02 in.)**	**1.00 mm (0.039 in.)**
0.33 mm (0.013 in.)	0.66 mm (0.026 in.)	1.32 mm (0.052 in.)

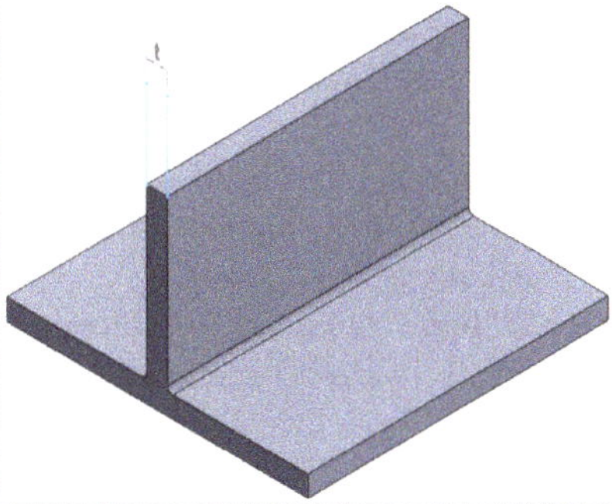

Comments

Warping may occur with extended lengths of unsupported walls (i.e., no ribs or intersecting walls). In this case, avoid using the minimum wall thickness.

Always test your printer to figure out which particular wall thicknesses leave a small gap between an inner and outer wall.

Avoid sharp transitions. Fillets at the points where walls join are recommended.

In general, an even wall thickness is recommended on all walls, both vertical and horizontal.

8.1.7 Feature Type: Horizontal Walls

With material extrusion technologies, horizontal walls can, theoretically be as thin as a single layer of material. In practice, however, to produce a horizontal wall with some strength and consistency, at least 4 layers of material are recommended.

Again, it is best practice to keep all walls of your product at the same thickness.

8.1.8 Feature Type: Support Material Overhang Angles

Maximum overhang angle (a)
45° This is a safe default number. But the angle can vary greatly from printer brand to printer brand, and depends on the desired surface quality

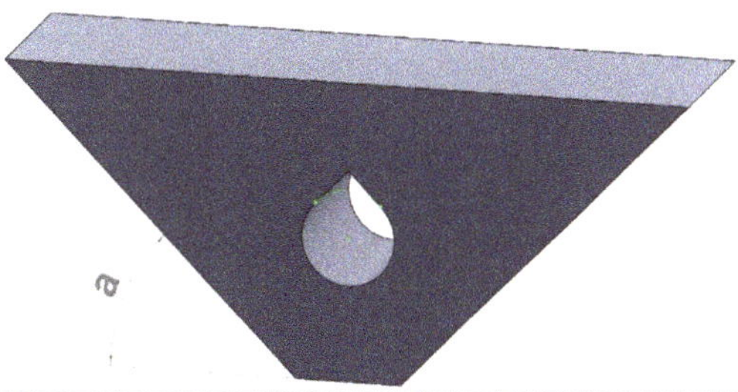

Comments

Overhang angles less than 45° (measured from horizontal) require support material, which is normally added automatically by the system software. Be aware that some systems measure the support angle from the horizontal, while others measure from the vertical.

Excessive supports that need to be broken away manually will increase post-processing time. Soluble support structures require much less manual labour, but they still waste material.

Horizontal holes (e.g., cooling channel profiles) can often be modified into teardrop or ovals shapes to minimize the need for internal supports that are hard to remove.

8.1.9 Feature Type: Clearances Between Moving Parts with Soluble Supports

Process variable	Minimum clearance	
Layer thickness	Horizontal (h)	Vertical (v)
0.18 mm (0.0071 in.)	0.36 mm (0.014 in.)	0.18 mm (0.0071 in.)
0.25 mm (0.0098 in.)	**0.50 mm (0.02 in.)**	**0.25 mm (0.0098 in.)**
0.33 mm (0.013 in.)	0.66 mm (0.026 in.)	0.33 mm (0.013 in.)

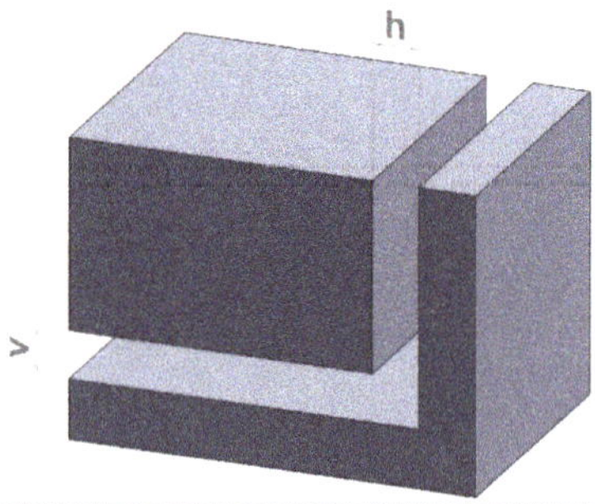

Comments

Large areas of close proximity will slow down the removal of support material. Clearance between parts built separately and assembled later must be at least equal to the general build tolerance of the system.

8.1.10 Feature Type: Clearance Between Moving Parts with Break-Away Support Material

Process variable	Minimum clearance	
Layer thickness	Horizontal (h)	Vertical (v)
0.18 mm (0.0071 in.)	0.36 mm (0.014 in.)	Adequate access to facilitate supports removal
0.25 mm (0.0098 in.)	**0.50 mm (0.02 in.)**	
0.33 mm (0.013 in.)	0.66 mm (0.026 in.)	

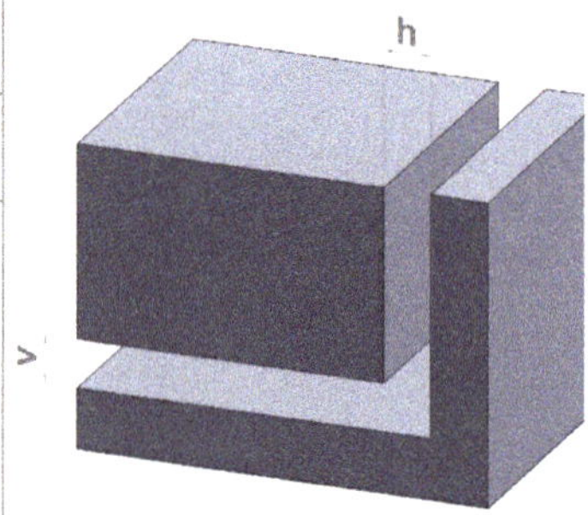

Comments

The main challenge with printing moving parts on a printer without soluble support is in the difficulty of removing the support material from between the moving parts.

Large areas of close proximity will slow down the removal of support material. Clearance between parts built separately and assembled later must be at least equal to the general build tolerance of the system.

8.1.11 Feature Type: Vertical Circular Holes

Required diameter (d)	CAD model diameter
5.0 mm (0.197 in.)	5.2 mm (0.205 in.)
10.0 mm (0.394 in.)	10.2 mm (0.402 in.)
15.0 mm (0.591 in.)	15.2 mm (0.598 in.)
20.0 mm (0.787 in.)	20.2 mm (0.795 in.)

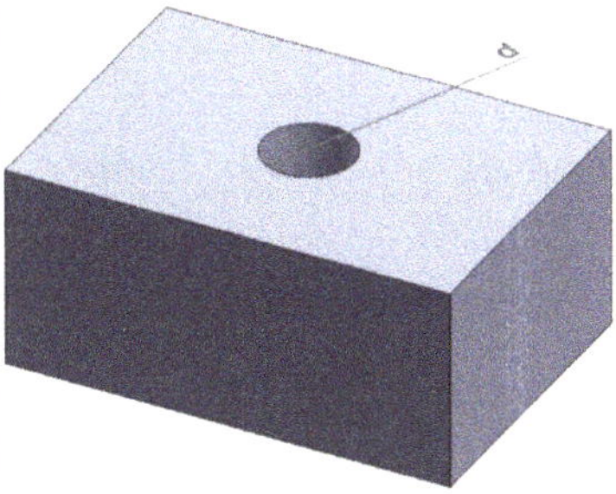

Comments

Holes are generally built undersized, typically by around 0.2 mm (0.0079 in.) across the diameter (Note: this value needs to be verified for each machine/material combination being used). This can be remedied approximately by adjusting the CAD model using the values above, or more precisely by drilling out the hole after the part has been built.

If using self-tapping screws, the column of contour material that surrounds the hole can sometimes be stripped out by the screw. A drop of super-glue, allowed to wick in between the contour and fill material, can alleviate this problem.

8.1.12 Feature Type: Circular Pins

Minimum diameter for vertical pins (v)	Minimum diameter for horizontal pins (h)
2.0 mm (0.079 in.)	2.0 mm (0.079 in.)

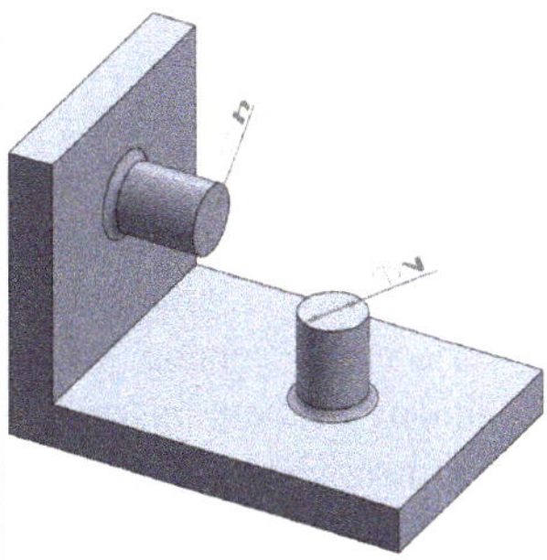

Comments

Pins with small diameters, vertical ones in particular, are prone to breaking off if only supported at one end.

Always fillet the pin where it joins the wall. Even 0.5 mm is enough to substantially strengthen the pin.

8.1.13 Feature Type: Built-in Screw Threads

Minimum thread diameter (d)	Minimum "dog-point" lead in (l)
5.0 mm (0.197 in.)	1.0 mm (0.039 in.)

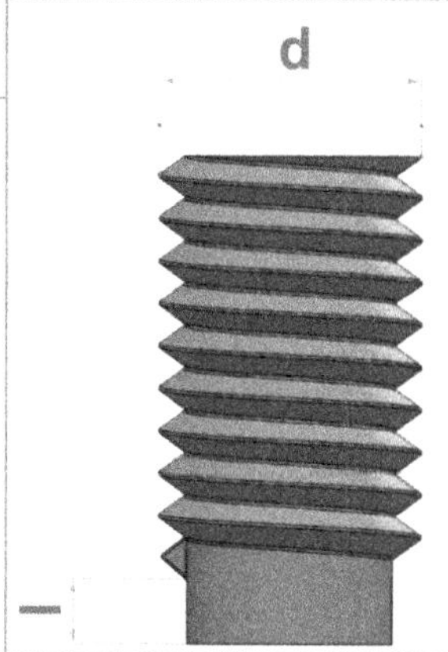

Comments

Use rounds on root and crest of thread. Tapping is recommended for small threads in holes and on posts.

Fillet the screw boss at the point where it meets the wall to avoid stress concentrations. A general guideline is to make the radius ¼ of the thickness of the wall.

8.2 Designing for Polymer Powder Bed Fusion

Polymer powder bed fusion (e.g., laser sintering, selective laser sintering) is an AM process in which thermal energy selectively fuses regions of a powder bed. The most common material used is polyamide (nylon), and there now exists a range of other polyamide based materials which include fillers such as glass, carbon, aluminium, etc. The part being constructed does not normally require supports, since the unfused powder surrounding the part provides sufficient support. This gives the designer greater freedom than most other AM system.

Parts created using polymer powder bed fusion usually have some degree of anisotropy in their material properties, particularly for small features that are less than about 25 mm^2 in surface area in the vertical direction, i.e., the part will be weaker or more brittle in the Z direction than it is in the X and Y directions. Designers must take this into account by, for example, ensuring that highly stressed features are built in a horizontal rather than vertical direction. This requires close cooperation with the AM process operator.

Another characteristic of this process is a pronounced granular roughness on the surface of parts. This can be reduced by various post-processing techniques (such as tumbling with abrasive media), but this will have an impact on part accuracy. Painting with automotive paints is also commonly used as a method for improving the surface finish and colour of polymer powder bed fusion parts, and this is discussed in greater detail in the post-processing chapter of this book.

8.2.1 Powder Bed Fusion Accuracy and Tolerances

There is a difference in the accuracy and tolerance between different manufacturers systems. They also vary depending on geometric features and print orientation. The only sure way to determine the accuracy and tolerance of any particular system is to print a test reference part and measure it. The numbers given below are for industrial quality powder bed fusion systems.

Layer thickness	0.1 mm (0.005 in.)
Accuracy	±0.3% lower limit of ±0.3 mm (0.010 in.)
Tolerance	±0.25 mm (0.010 in.) or ±0.0015 mm/mm (0.0015 in./in.)—whichever is greater
Smallest feature size	Around 0.5 mm (0.04 in.)

8.2.2 Layer Thickness

The typical layer thickness for powder bed fusion is 0.1 mm, but some systems allow for layer thicknesses of 0.06 mm. Compared to other AM technologies, however, the stair-stepping effect is less visible on polymer powder bed fusion technologies. It is only on very gently curving surfaces of a relatively large surface area that it is visible.

8.2.3 Avoiding Large Masses of Material

As with injection molding, designers need to be careful with trying to avoid uneven thicknesses of plastic in their products and, in particular, avoid large masses of material. These can both cause distortion in the part, and can also add substantially to the time it takes to make the part, because of the laser having to 'hatch' all the extra surface area caused by the large mass, and therefore its cost. Refer to the chapter on the economics of AM for more details.

The most common technique for removing large masses of material is to shell the part. Then a decision needs to be made about whether or not to keep the loose powder hat is trapped inside the part. If the part benefits from being lighter, then the simple addition of some salt-shaker holes to the part will allow the loose powder inside the part to be removed.

8.2.4 Powder Age and Refresh Rate

Many polymer powder bed fusion systems use a mix of virgin powder and used powder. Typically, the ratio of new to old powder ranges from 20/80 to 35/65. There is, however, a limit as to how many times powder can be reused before it begins to produce undesirable parts. As powder ages printed parts can develop an 'orange peel' effect in which there outer surface is mottled with little dimples. When parts reach this state, it is best to safely dispose of the powder and begin again with a mix of virgin and slightly used powder.

The following pages contain information on how to design specific features to be built using the polymer powder bed fusion process.

8.2.5 Feature Type: Wall Thickness

Minimum wall thickness (t)	Recommended minimum wall thickness (t)	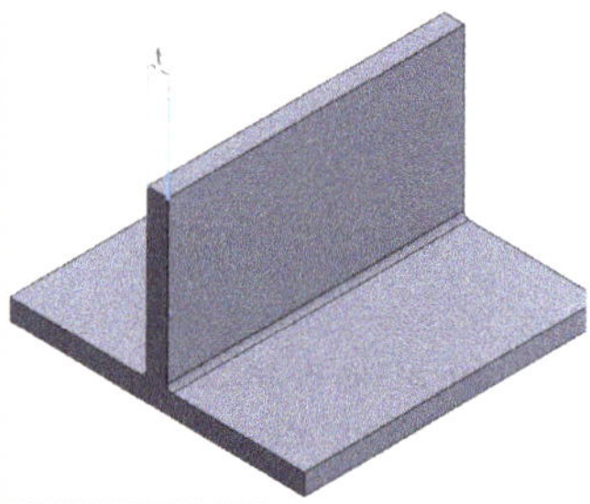
0.6–0.8 mm (0.031 in.)	1.0 mm (0.039 in.)	

Comments

Though it is, on occasion, possible to print walls thinner than the 0.6 mm number given above, their success is highly dependent on the rest of the part geometry, print orientation, etc.

Thin walls with a large surface area are likely to warp during the cooling process. If large surface area thin walls are required, consider adding ribs to stiffen the walls.

Thicker walls and any large volumes of material will result in excess heat retention in the part and hence shrinkage, resulting in geometric deformation. Therefore, a maximum wall thickness of 1.5–3 mm is also recommended. If walls must be thicker than this, consider shelling them. This will both help to reduce distortion and greatly speed up print times.

In general, an even wall thickness is recommended on all walls, both vertical and horizontal.

8.2.6 Feature Type: Clearance Between Moving Parts

Minimum horizontal clearance (h)	Minimum vertical clearance (v)
0.5 mm (0.02 in.)	0.5 mm (0.02 in.)

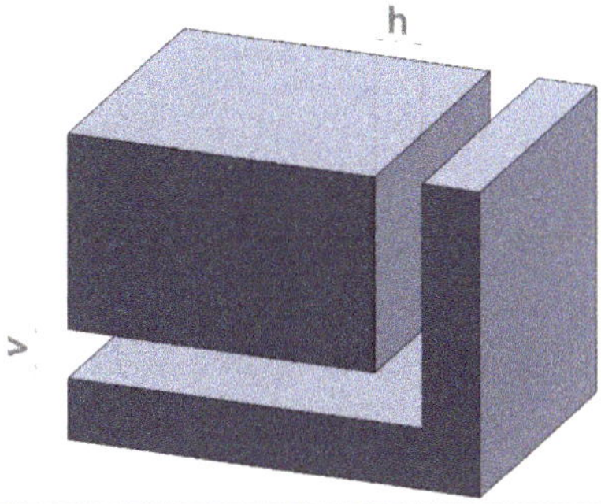

Comments

The required gap between moving parts is highly dependent on the surface area of the faces that are in close proximity. If faces in close proximity have surface areas of only a few mm^2, then gaps as small as 0.2 mm between the faces are possible. The 0.5 mm gap stated above is one that will work in most situations and on most different manufacturers systems.

Large areas of close proximity will slow down the removal of excess powder. Clearance between parts built separately and assembled later should be at least equal to the general build tolerance of the system.

8.2.7 Feature Type: Circular Profile Through Holes

Process variable	Minimum diameter	
Wall thickness	Vertical hole (v)	Horizontal hole (h)
1 mm (0.039 in.)	0.5 mm (0.019 in.)	1.3 mm (0.051 in.)
4 mm (0.157 in.)	0.8 mm (0.031 in.)	1.75 mm (0.069 in.)
8 mm (0.314 in.)	1.5 mm (0.059 in.)	2.0 mm (0.079 in.)

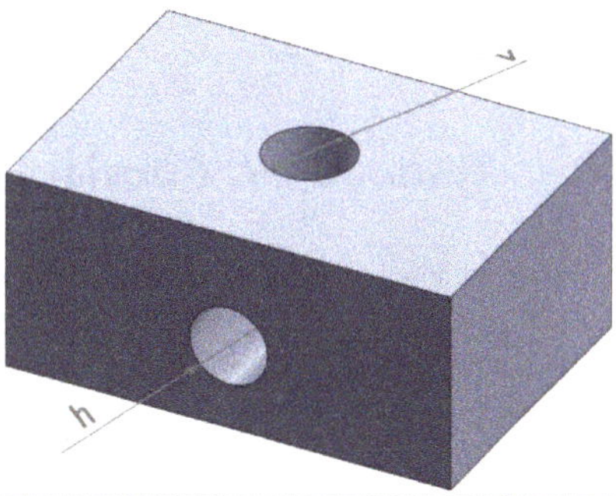

Comments

Small holes, both round and square, typically below 1.5 mm are related closely to wall thickness. As the wall thickness increases, powder becomes increasingly difficult to clear from small holes. As the wall thickness decreases, smaller through holes become feasible.

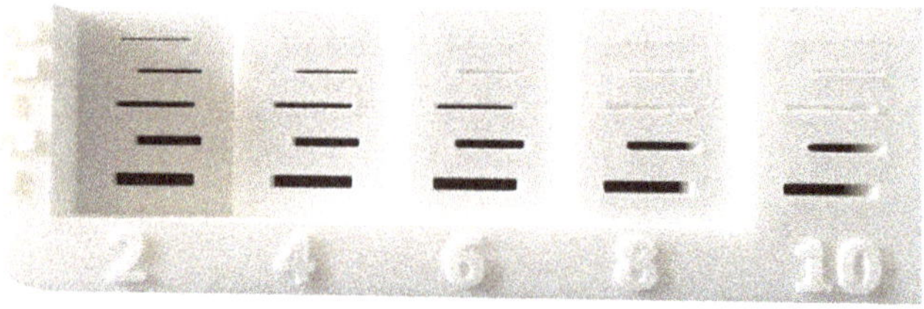

8.2.8 *Feature Type: Square Profile Through Holes*

Process variable	Minimum diameter	
Wall thickness	Vertical hole (v)	Horizontal hole (h)
1 mm (0.039 in.)	0.5 mm (0.019 in.)	0.8 mm (0.031 in.)
4 mm (0.157 in.)	0.8 mm (0.031 in.)	1.2 mm (0.047 in.)
8 mm (0.314 in.)	1.5 mm (0.059 in.)	1.3 mm (0.051 in.)

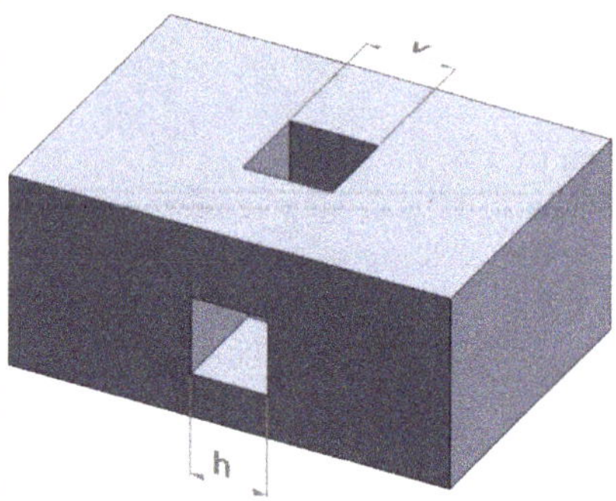

Comments

Small holes, both round and square, typically below 1.5 mm are related closely to wall thickness. As the wall thickness increases, powder becomes increasingly difficult to clear from small holes. As the wall thickness decreases, smaller through holes become feasible.

8.2.9 *Feature Type: Circular Pins*

Minimum diameter for vertical pins (v)	Minimum diameter for horizontal pins (h)
0.8 mm (0.031 in.)	0.8 mm (0.031 in.)

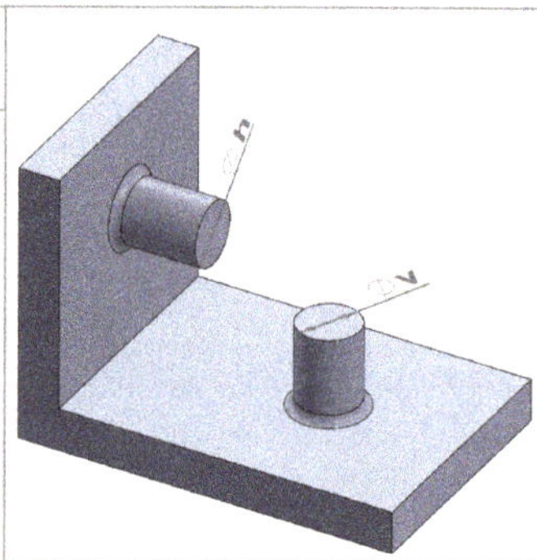

Comments

Pins with small diameters are prone to breaking off if only supported at one end.

Always fillet the edge where a pin joins a face.

8.2.10 Feature Type: Hole Proximity to Wall Edge

Design variable	Minimum distance to edge	
Hole diameter	Vertical hole (v)	Horizontal hole (h)
2.5 mm (0.098 in.)	0.8 mm (0.031 in.)	0.8 mm (0.031 in.)
5.0 mm (0.197 in.)	0.9 mm (0.035 in.)	0.95 mm (0.037 in.)
10.0 mm (0.394 in.)	1.05 mm (0.041 in.)	1.0 mm (0.039 in.)

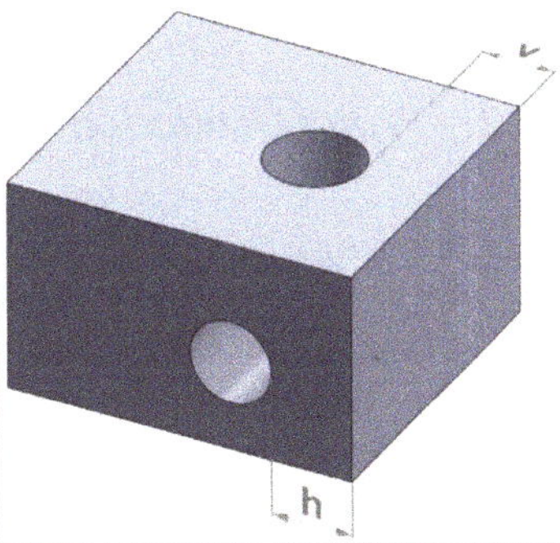

Comments

Larger holes require slightly greater distances to the edges of walls.

8.3 Designing for Vat Photopolymerisation

A vat photopolymerisation machine, aka SLA, will contain a UV laser to cure a specific layer of a component from a tank of photosensitive resin. Though many of the other polymer design rules apply equally to SLA, there are certain guidelines that are specific to resin based technologies.

Note that some SLA systems will fabricate the part top-down, and gradually pull the part down into the vat of resin as it is being built, while others, desktop SLA systems in particular, will tend to work bottom-up, where the laser source (or DLP/DMD) is below the vat, and solidifies the part through a window at the bottom of the vat, and pulls the part out of the resin vat.

8.3.1 Resolution

SLA is able to achieve a higher resolutions than some other additive manufacturing processes. SLA resolution in the XY-direction is dependent on the laser spot size and can range anywhere from 50 to 200 μm. This is not usually an adjustable parameter of printing. The minimum feature size can therefore not be smaller than the laser spot size.

Resolution in the Z-direction varies from 25 to 200 μm depending on the choices of layer thickness allowed by the machine. Choosing a very fine vertical resolution is a trade-off between speed and quality. For a part that has few curves or fine details, there will be little visual difference between a print at 25 μm compared to a print at 100 μm.

8.3.2 Print Orientation

When orienting a part for SLA, particularly on a bottom-up SLA machine that cures the part from the bottom and pulls it out of the vat of resin, the biggest concern is vertical cross-sectional area. The forces involved with a print sticking to the bottom of the tank are directly proportional to the 2D cross-sectional area of the print. Because of this, a part with large cross-sectional areas, is often best printed at an angle to the plate. Minimizing the cross-sectional area along the Z-axis is the best way to orient parts for SLA prints.

The part design can, of course, also be modified to minimize large cross-sectional areas. It is important to understand why part orientation affects the quality of an SLA print. The necessity to orient components so that the Z-axis cross-sectional area is minimized often results in a significant amount of support material being needed. In some cases, a design may need so much support that printing in SLA is no longer cost effective, or so detrimental to the surface quality of the part that the end result is no longer of good enough quality to be used.

Reducing the number of horizontal areas relative to the print orientation, hollowing out components and reducing the cross-sectional area are all steps that can be taken to optimize a design for SLA.

8.3.3 Support Material

SLA does require support material for overhanging features. This is because the uncured resin is not viscous enough to support features on its own. This support material must be removed in post-processing. On most SLA system the process of adding support material to the part is largely automated but, with experience, the user can manually edit the supports to avoid having supports in areas where surface finish is critical (Fig. 8.5).

8.3.4 Overhangs

Overhangs generally pose very little issue with SLA, unless the model is being printed without adequate support structures. Printing without supports often leads to

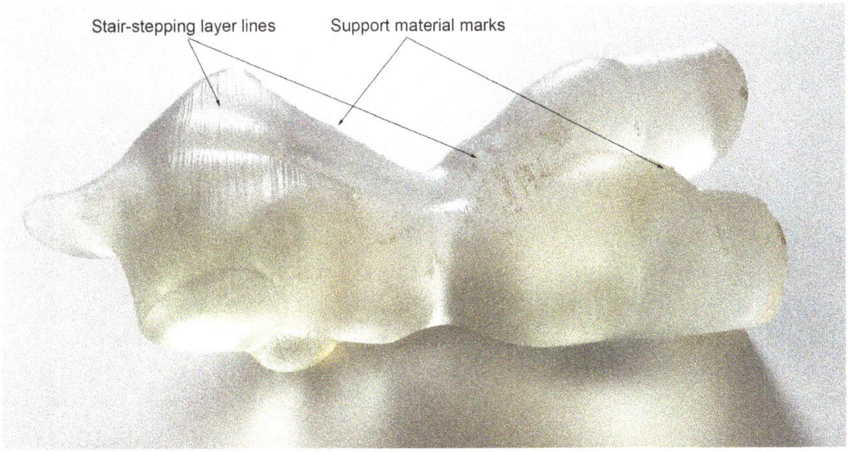

Fig. 8.5 Examples of marks left by support material and of the stair-stepping effect caused by the layers of materials

warping of the print, but if printing without supports is necessary, any unsupported overhangs must be kept less than 1.0 mm in length and at least 20° from horizontal.

8.3.5 Isotropy

SLA is one of the few processes where the parts are relatively isotropic. This is because the layers chemically bond to one another as they print, resulting in near identical physical properties in the X, Y, and Z-direction. Whether the part is printed horizontally, vertically, or at an angle to the build plate, the material properties of the part will not be noticeably different in any particular direction.

8.3.6 Hollowing Parts and Resin Removal

SLA machines can print solid, dense models but, if the print is not intended to be a functional part, shelling the model to be hollow can significantly reduce the amount of material needed as well as reduce the print time. It is recommended that the walls of the hollowed part be at least 2 mm thick to reduce the risk of failure during printing.

If printing a hollow part, drainage holes must be added to allow the uncured resin to be removed from the part. If left inside the part, the uncured resin can create a pressure difference within the hollow chamber, and can cause what is known as "cupping". Small failures such as cracks or holes propagate throughout the part and will, eventually, cause complete failure, or part explosion, if not corrected.

Drain holes should be at least 3.5 mm in diameter, and at least one hole must be included per hollow section, although two holes can make the resin much easier to remove.

8.3.7 Details

Embossed details (including text) include any features on the model that are raised slightly above the surfaces around them. These must be at least 0.1 mm in height above the surface of the print to ensure the details will be visible.

Engraved details (including text) include any features which are recessed into the model. These details are at risk of fusing with the rest of the model while printing if they are too small, so these details must be at least 0.4 mm wide and at least 0.4 mm deep.

8.3.8 Horizontal Bridges

Bridges between two points on a model can be successfully printed, but one must keep in mind that wider bridges must be kept shorter (usually under 20 mm) than thin bridges. Wider bridges have a greater cross-sectional area of contact which increases the chance of print failure during delamination from the bottom window.

8.3.9 Connections

If parts are being made that need to connect together, it is always best have a certain tolerance between the parts that fit together. For SLA, these tolerances are:

- 0.2 mm clearance for assembly connections.
- 0.1 mm clearance will give a good push or snug fit.

If interlocked moving parts are being printed, then the tolerance should be 0.5 mm between the moving parts.

8.3.10 Feature Type: Wall Thickness

Minimum wall thickness for supported walls (t)	Minimum wall thickness for unsupported walls (t)	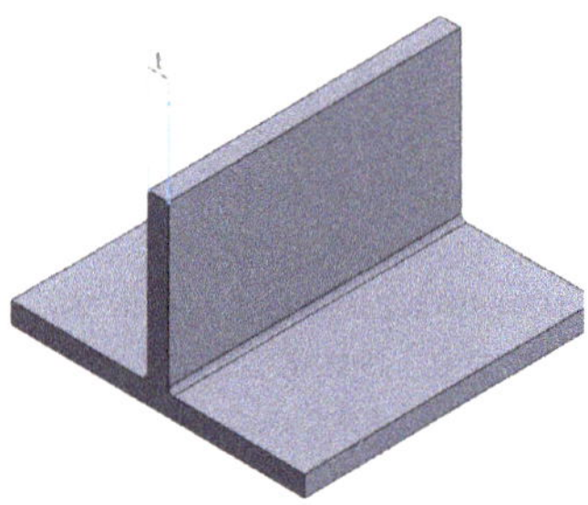
0.4 mm (0.016 in.)	0.6 mm (0.023 in.)	

Comments

Supported walls are walls that are connected to other structures on at least two sides, so they have very little chance of warping. These should be designed at a minimum of 0.4 mm thick. Note that if the supported wall has a large surface area, a larger thickness may be required.

Unsupported walls are walls that are connected to the rest of the print on less than two sides, and are at a very high chance for warping or detaching from the print. These walls must be at least 0.6 mm thick.

Always fillet the corners where one wall meets another wall to reduce stress concentrations along the joint.

In general, an even wall thickness is recommended on all walls, both vertical and horizontal.

8.3.11 Feature Type: Circular Holes

Minimum diameter h and v	
0.5 mm (0.019 in.)	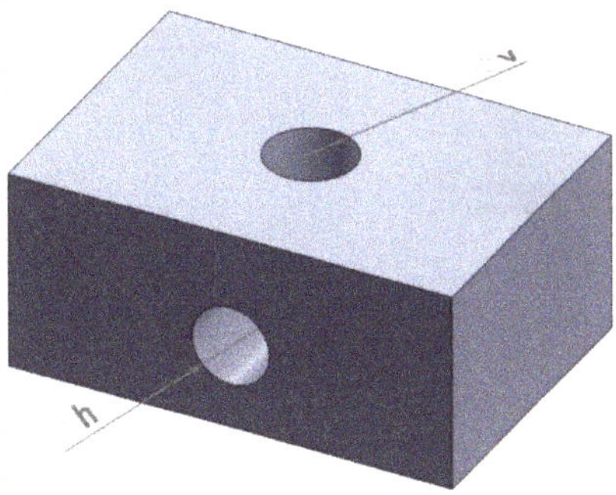

Comments

Holes with a diameter less than 0.5 mm in the X, Y, and Z axes may close off during printing.

Chapter 9
Design for Metal AM

9.1 Designing for Metal Powder Bed Fusion

Metal powder bed fusion (e.g., direct metal laser sintering, selective laser melting, and electron beam melting) is an AM process in which thermal energy selectively fuses regions of a powder bed. Materials used include stainless steel, tool steel, aluminium, titanium alloys, nickel-based alloys, cobalt chrome, and precious metals such as gold. The part being constructed normally requires supports (sometimes called anchors) to be added, built from the same material as the part. These supports are removed manually after the build process, so the designer must allow room for access to the supports. Also, care must be taken when creating small features that are attached to the supports because they can break off accidentally during support removal.

One aspect that has been, somewhat, miscommunicated by both media and machine manufacturers is that metal AM is easy. This is most certainly not the case, and an understanding of design for metal AM is essential to be able to produce cost-effective and value adding parts. Understanding that printing is only a very small part of the metal AM part production process, and gaining an appreciation of both the pre- and post-processing that will be required after the part has been printed allows engineers to use AM to add true value to their products.

Metal AM technologies can, broadly, be broken down into the categories shown below. In this chapter, however, we focus mainly on designing for powder bed fusion, because it is the process that is currently the most utilized by industries in different application sectors. A later chapter also discusses design aspects for the metal binder jetting process. No doubt, in the future, as other technologies become more widely adopted, we will need to expand the design methods for those technologies. Future editions of this book will cover such design methods as they are developed (Fig. 9.1).

O. Diegel et al., *A Practical Guide to Design for Additive Manufacturing*, Springer Series in Advanced Manufacturing,
https://doi.org/10.1007/978-981-13-8281-9_9

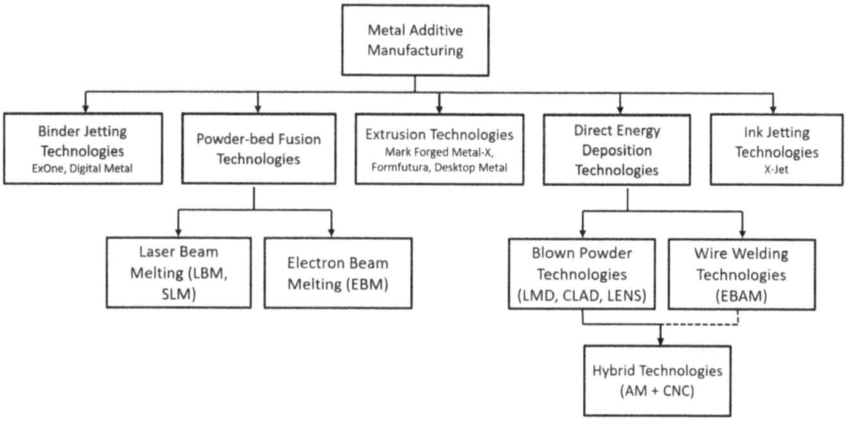

Fig. 9.1 Hierarchy of metal AM technologies

9.2 The Basics of Powder Bed Fusion

In order to understand powder bed fusion technologies, one must gain an understanding of both the powders that are used in these technologies, as well as the characteristics of the powder fusing process itself. Without this fundamental understanding, it may be hard to understand why and how defects may have occurred in a metal AM part.

9.3 Metal Powder Production

Metal AM powder is, most commonly, made through a gas atomization process. There are a number of different atomisation processes including gas atomisation, vacuum induction melting gas atomisation, plasma atomisation, centrifugal atomisation, and water atomisation. Most of these atomization processes produce:

- A spherical powder shape
- A good powder density, thanks to the spherical shape and particle size distribution
- A good reproducibility of particle size distribution

Gas atomisation, one of the common powder producing processes, works by blasting a stream of molten metal with a neutral gas jet, which blasts the metal into spherical particles (Fig. 9.2).

Vacuum induction melting (VIM) gas atomisation is similar, but has the whole process occurring in a vacuum, which can help to reduce the oxidisation that may occur in the metal powder (Fig. 9.3).

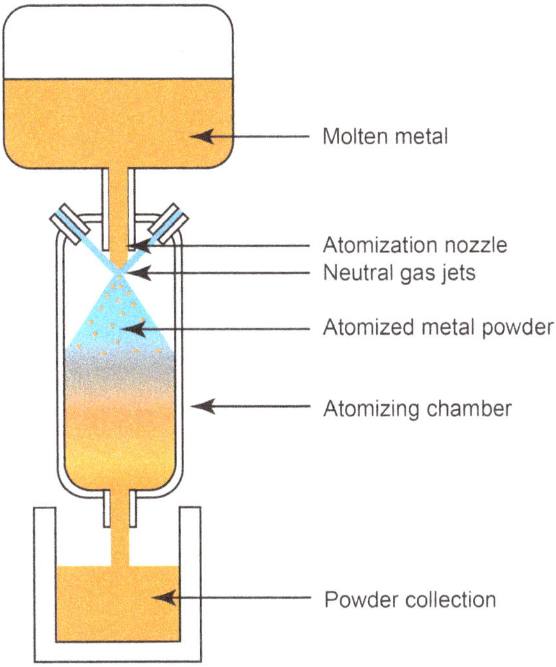

Fig. 9.2 Schematic diagram of a gas atomisation system

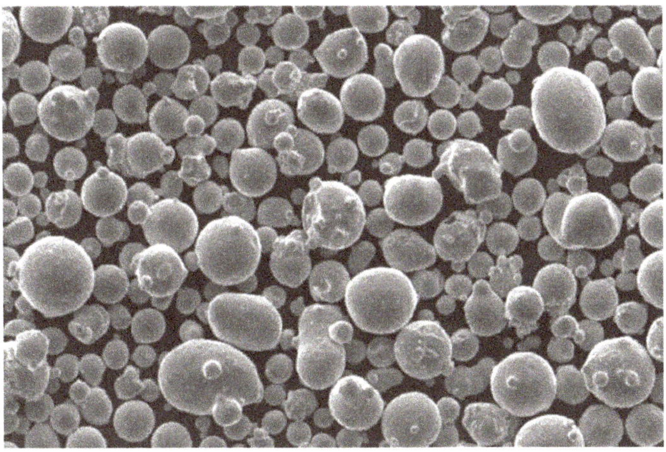

Fig. 9.3 Powder produced by vacuum induction melting (Pearl® Micro Ni718). Courtesy of Aubert & Duval

9.4 Powder Morphology (Ideal Powder Shape)

The recommended powder shape for metal AM is spherical because it is beneficial for powder flowability and helps to form uniform powder layers in powder bed systems. All powder bed fusion systems work by spreading thin layers of powder before selectively fusing each layer, so the better the powder can be spread (smooth layers as densely packed as possible, and without any 'streaking'), the better the part quality will be.

Typical defects to be controlled and minimized in powder are:

- Irregular powder shapes such as elongated particles, which may cause the powder to be harder to spread evenly.
- Satellites which are small powder grains stuck on the surface of bigger grains, which will make it harder to spread or leave 'streaks' in the layer.
- Hollow powder particles, with open or closed porosity. These can explode during the melting process, or can entrap gas in the part which may produce parts with undesirable results (Fig. 9.4).

9.5 Powder Size Distribution

For powder-bed fusion the powder size most commonly used is between 30 and 40 μm, with a bell-curve distribution with some large and some smaller particles (Fig. 9.5). Some systems that allow for very thin layer thicknesses may require smaller particle sizes. Some materials, such as aluminium, for example, may have a slightly larger powder size distribution than, say, steel or titanium.

Larger powder sizes of between 50 and 100–150 μm are commonly used for EBM and DED technologies.

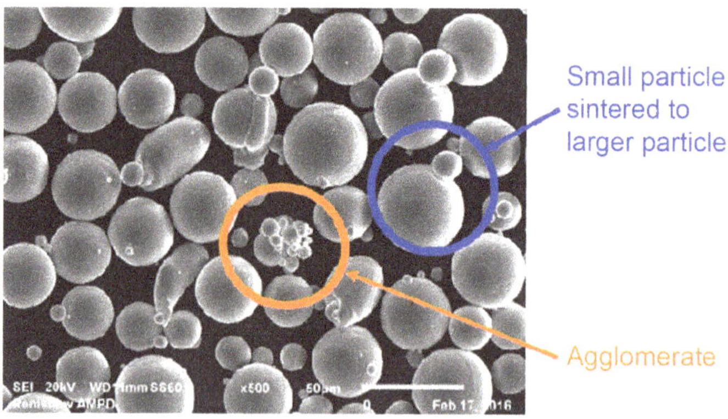

Fig. 9.4 Undesirable powder defects. Courtesy of Renishaw

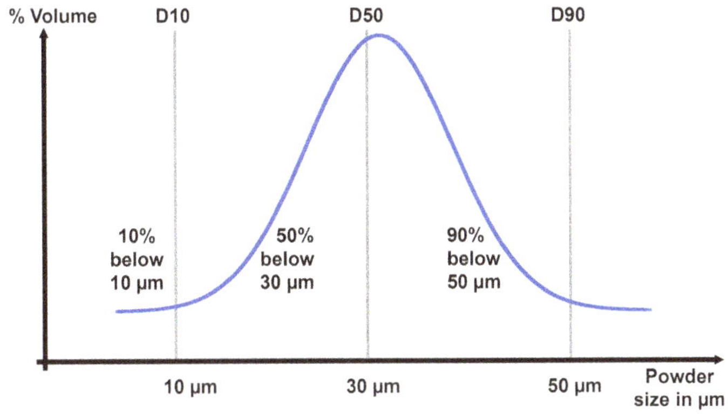

Fig. 9.5 Typical powder size distribution for laser based powder bed fusion

The reason a mixed powder size distribution is desirable is so that the smaller particles fit between the larger ones, and allow a denser layer of powder to be spread. If all the particles are exactly of the same size, this will leave gaps between the spread powder particles, which will cause it to collapse, or shrink, more during the meting process.

9.6 Other Powder Considerations

Metal powders can be hazardous to the health if not handled properly. They can also cause explosions. These are discussed further in the chapter on AM health and safety.

Some of the other factors to consider when dealing with metal AM powders include:

- Powder storage, handling and aging: For almost all alloys, shielding gas, the control of hygrometry and temperature is important and strongly recommended.
- Powder reusability: i.e. the definition of conditions of re-use of unused powders after additive manufacturing cycles (sieving of agglomerates, control, number of re-use etc.).
- Health, safety and environmental issues.

9.7 Metal AM Material Characteristics

The precise metallurgical and mechanical properties of metal AM parts vary vastly from system manufacturer and machine to system manufacturer and machine, and material to material. This is because each system manufacturer uses their own laser parameters and scan strategies. In addition, many users may further modify the laser parameters and scan strategies to suit their own particular needs. In addition, there are, literally, hundreds of factors that have an impact on the mechanical properties of AM parts. This is one of the reasons that, when reading different publications about metal AM about experiments in which the same material is being used on the same brand of machine, each publication may produce different metallurgical or mechanical properties.

Some work undertaken by Dr. Christoph Haberland [3], of Siemens Turbomachinery, created an Ishikawa diagram of the various factors that affect AM part quality. This image, in its own right, paints a true picture of the sheer complexity of the many parameters that can have an effect on producing good metal AM parts (Fig. 9.6).

In general, however, if we compare the characteristics of metal AM parts to conventionally produced metal parts, AM parts, straight off the machine, with no post-processing other than support material removal and shot-peening, would perform as follows:

	Mechanical properties	Surface finish
Sand cast	AM superior	AM superior
Investment cast	AM superior	AM inferior
Wrought or forged	AM inferior	AM inferior

With suitable post-processing, however, AM parts can, in some cases, approach the mechanical properties of forged or wrought parts.

If we then look a bit more deeply at the metallurgical characteristics of metal AM parts, some general characteristics of materials produced by AM would include:

- A fine microstructure, due to the very rapid solidification process.
- A slight anisotropy in Z direction, so slightly different mechanical properties in the Z direction. Note that these different properties are not necessarily a weakness.
- A few small residual porosities, in particular below the surface. But densities of 99.9% are commonly reached with AM processes.
- To achieve full density, and reduce anisotropy, Hot Isostatic Pressing (HIP) is commonly used (Fig. 9.7).

9.8 Potential Defects in AM Materials

In the case of incorrect process parameters, build strategy, part orientation or insufficient powder quality, some typical defects that can be observed in metal AM parts include (Fig. 9.8):

- Unmolten powder particles.
- Lacks of fusion.
- Pores.
- Cracks.
- Inclusions.
- Residual stresses.
- Poor surface roughness.

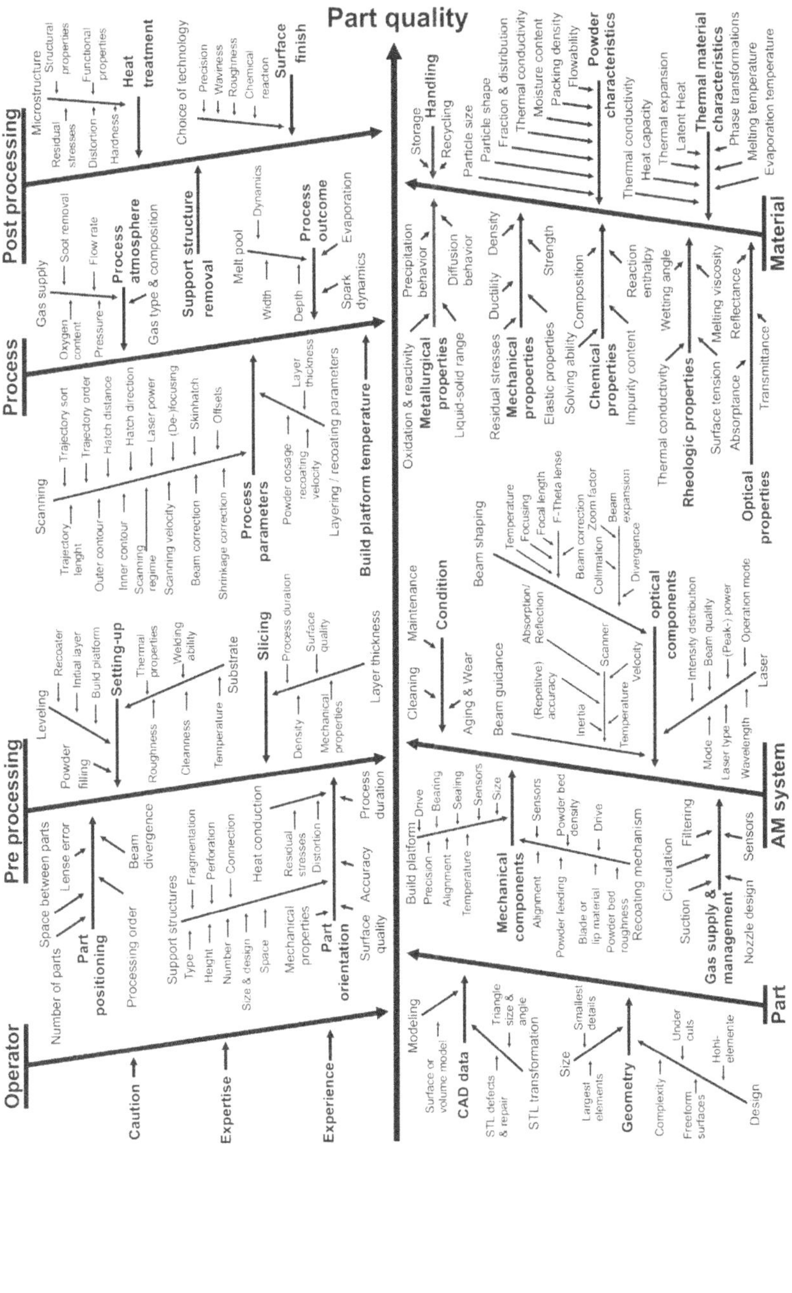

Fig. 9.6 Ishikawa diagram of parameters that affect metal AM parts. Courtesy of Dr. Christoph Haberland, Siemens [3]

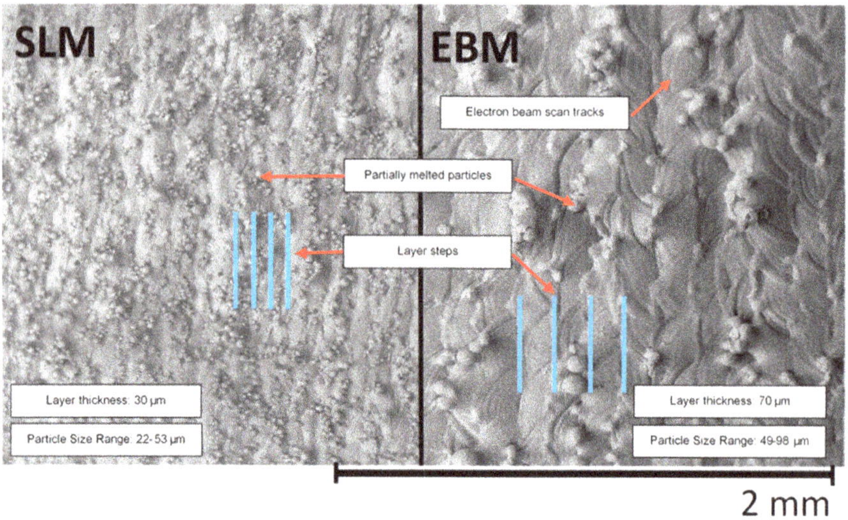

Fig. 9.7 Metallurgical properties of metal AM parts, Andrew Triantaphyllou, MTC

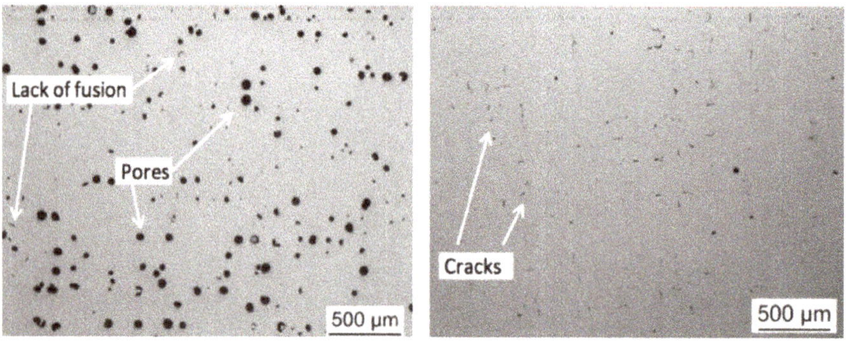

Fig. 9.8 Potential defects in metal AM pats. Courtesy of IK4 Lortek

9.9 The Metal AM Process

The powder bed fusion process begins by spreading a thin layer of powder onto the build plate, and an energy source is then used to scan the powder and fuse it together where required. The process is then repeated for subsequent layers. What is happening during the fusing part of each slice is illustrated below (Fig. 9.9).

The energy from the laser, or electron beam, has the effect of fusing the powder together. Not all the energy, however, is absorbed by the powder. Some is scattered, and some is reflected (Fig. 9.10).

There are several parameters that the machine operator has control over that will affect the quality of the part. Some of them, such as the beam offset, are set during

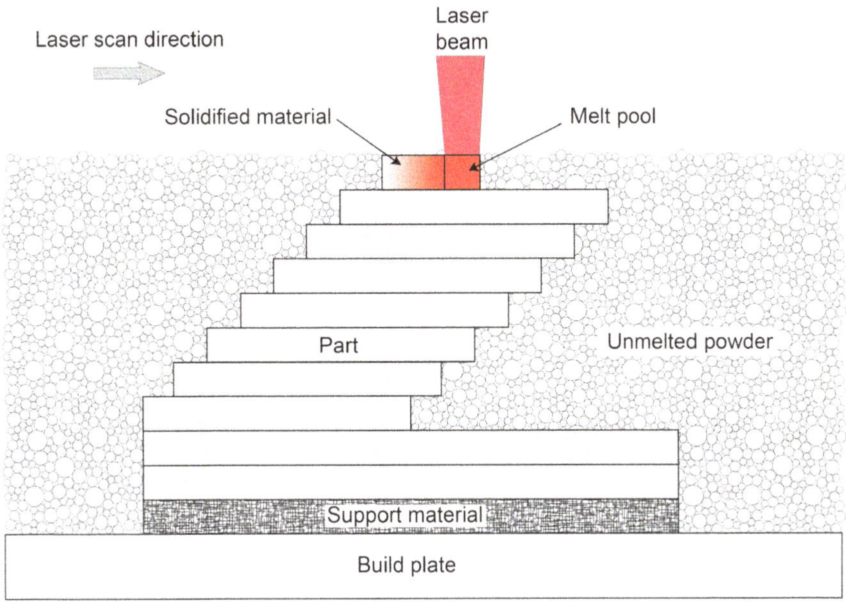

Fig. 9.9 The overall metal AM build process

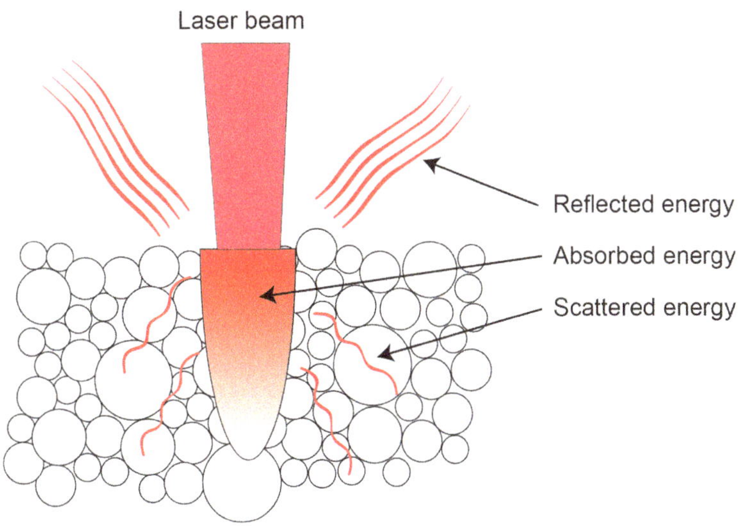

Fig. 9.10 The effect of powder on energy beam absorption

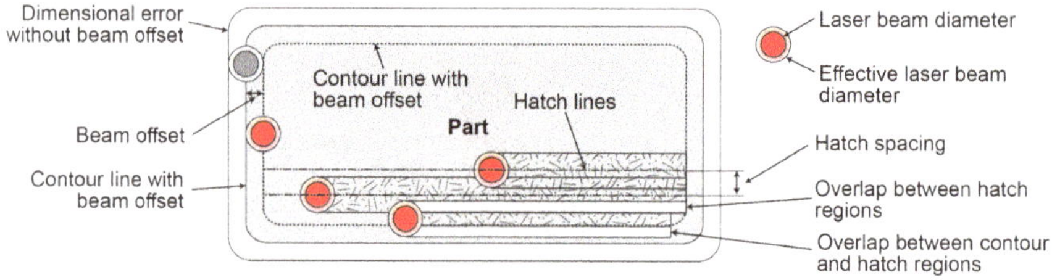

Fig. 9.11 Contour lines, beam offset and hatch spacing parameters

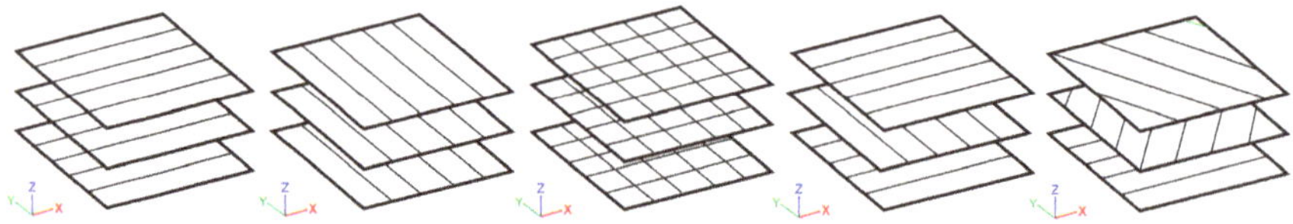

Fig. 9.12 Examples of different energy beam scanning strategies

machine calibration, and may then not be touched again until it is time to recalibrate the machine. But others, such as hatch spacing, are critical in thc production of the parts, as this affects both the quality of the part, and the speed at which it can be built. The hatch spacing is also one of the parameters that can be used to affect the density, or controlled porosity, of the part.

Note that not all machines allow all parameters to be controlled by the user, and that some may require additional payments to purchase an 'open' version of the software that allows access to the various print parameters (Fig. 9.11).

There are also various scanning strategies that can be used in order to minimize the stress in each layer of the part so as to minimize part distortion. A common scanning strategy, for example, is to rotate the scan pattern for each successive layer by 67°. This avoids consecutive layers having exactly overlapping scan patterns, which could increase residual stress in the part. Some scan strategies are discussed further in the section on minimizing residual stresses (Fig. 9.12).

In addition, there are different scanning patterns used for the energy beam to scan each slice of the model. Some can the entire slice in a continuous meander pattern, while others use a stripe, or chessboard, or hexagon pattern. These strategies will also be revisited in the chapter on minimizing residual stress.

9.9.1 Energy Density

Ultimately, it is the energy density that determines whether or not the powder fuses. Sufficient energy density is needed to melt powder particles of the layer being processed and of the previous layer to ensure strong joining between successive layers and avoid a lack of fusion and porosity. Excessive energy can cause vaporization

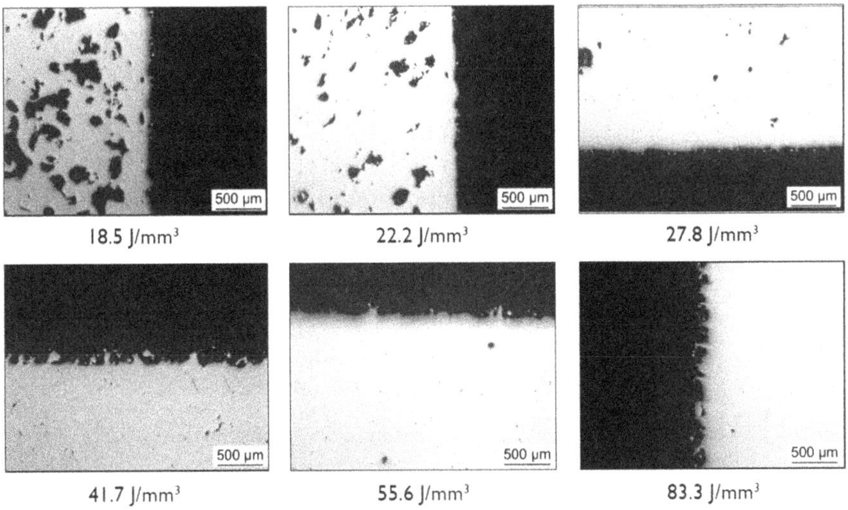

Fig. 9.13 The effect of energy density on surface roughness. Courtesy of IK4 Lortek

of the material, creating defects and reducing material density. Getting this energy density right is a delicate balance, and is determined by the formula below. As can be seen, all these parameters can easily be adjusted (assuming you are running 'open' machine software) by the user.

$$E = \frac{P}{v \cdot h \cdot t} \qquad \begin{aligned} &\mathrm{E} = \text{energy density} \\ &\mathrm{P} = \text{Power (W)} \\ &\mathrm{v} = \text{Scanning speed (mm/s)} \\ &\mathrm{h} = \text{hatch spacing (mm)} \\ &\mathrm{t} = \text{layer thickness (mm)} \end{aligned}$$

As with everything else in AM, the best energy density to use is somewhat of a compromise. As the energy density increases, the part density increases but, at a certain point, the surface finish begins to deteriorate. At higher energy densities, porosity is minimal near the surface, but roughness gets worse.

In the example below, using Ti6Al4 V as the material, energy densities above 40 J/mm^3 are needed to obtain parts with 99.7–99.9% relative density. As the energy density increases beyond that, part density continues to improve but surface roughness gets worse. This is because the increased energy causes the molten material to be violently agitated, which results in a rougher surface. At an energy density of 30 J/mm^3, however, the part is slightly less dense (but generally still better than 99% dense) but the part has both improved surface quality and minimized defects at the borders (Fig. 9.13).

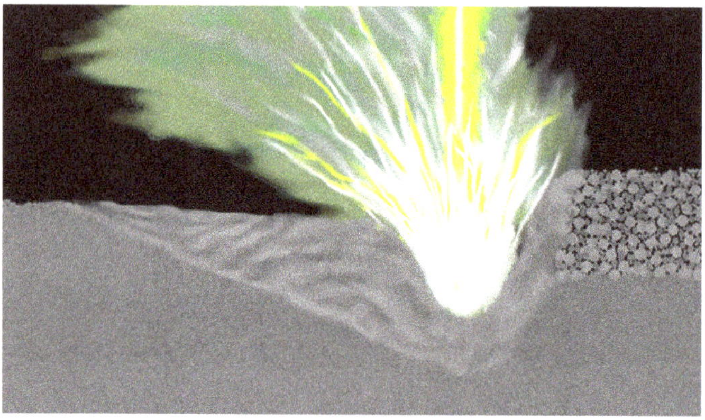

Fig. 9.14 The controlled chaos of additive manufacturing. Courtesy of Stratonics

Remember, also, that typical densities for cast components are often in the 98% range (depending on the material and casting process), so even a relatively low AM density may still be better than a cast density.

9.10 Controlled Chaos

The process of producing metal AM part can best be described as controlled chaos. Tiny particles of powder are being blasted by a moving energy beam, which causes some particles to fuse, others to fly out of the way, and some very small particles to be completely vaporised.

Yet, somewhat amazingly, by carefully balancing all the parameters we can control, we can produce good metal parts with near full density and relatively little (or well controlled) anisotropy (Fig. 9.14).

It is because of the many interacting parameters that AM is both extremely capable while, at the same time, being extremely complex. Understanding which parameters can be controlled within the machine itself, and which parameters are influenced by the design of a part becomes a critical issue in determining the success of metal AM parts.

9.11 The Reality of Metal AM

As metal AM is an expensive process (see the chapter on the economics of AM for more), and parts can require substantial post-processing, you need to have a good reason to make a metal AM part. Typically, parts that are not specifically designed for metal AM are not worth doing with AM. There are some exceptions to this, like

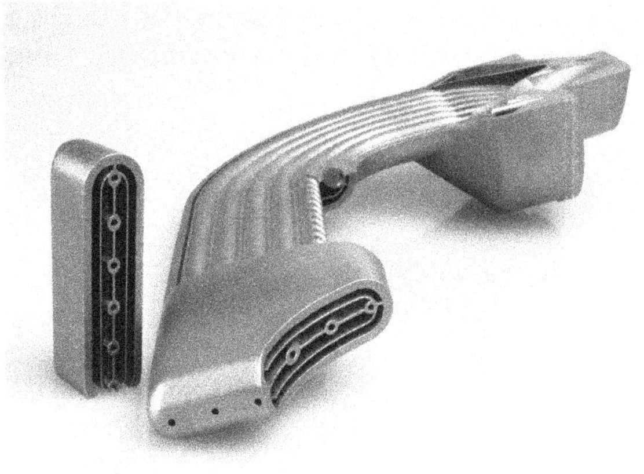

Fig. 9.15 A gas emissions rake optimized for AM. *Courtesy* RSC Engineering GmbH

spare parts for example, but, in general, the geometry of the part must be complex enough that it cannot easily be made through traditional manufacturing.

The gas emission rake, shown below, is an excellent example of when metal AM can add value where conventional manufacturing cannot. Its geometry is such that it could not be produced with conventional manufacturing, and being able to make it as a single metal AM component adds considerable value to the part (Fig. 9.15).

As most metal AM requires support structures (also referred to anchors) for heat transfer, and they can be difficult to remove, particular effort needs to put into design techniques that help to minimize the amount of support material used.

Part orientation is also of critical importance with metal AM, both for minimizing supports, and for surface finish.

In short, despite what the media and system manufacturers may claim, metal AM is hard work! It requires a great deal of knowledge both from the design engineer and the machine operator. The following sections of this book attempt impart some of that knowledge and design guidance to help the user to achieve successful metal AM parts.

9.12 Topology Optimisation

Because of the high costs associated with metal AM, topology optimisation is an excellent technique to apply to a number of metal parts. Because one of the benefits of AM is its ability to make very complex geometries, topology optimisation, which can often create cell-like biologically inspired structures, now offers the ability to make much lighter components than with conventional manufacturing. For any application where weight savings is a benefit (aeronautics, automotive, etc.) this can add great value to products.

Please refer to the chapter on topology optimisation for further information on how to apply this technique to additive manufacturing.

9.13 Lattice Structures

Lattices are another excellent way of producing light-weight, but extremely strong, parts that can also drastically reduce the time and cost it takes to make metal AM parts. Lattice structures offer a major advantage in reducing part weight without greatly affecting part strength, which is very important in industries such as aerospace and transportation.

A lattice is a cellular structure made of repeated unit cells to form a larger volume. There are many options for the shape and size of such lattice cells, and for the pattern in which they are is repeated, and there are countless examples of lattice structures being used to reduce the amount of material used in a part to improve its strength-to-weight ratio, or to replace support material in a part.

Lattices can be uniform, where the same cell size is repeated in all directions of the part, or variable, where the size and spacing of the cells is different in different directions. Variable lattice structures are popular in medical applications, such as implants, because the variability of the cells closely mimics the structure of our bones. This can provide comparable strength-to-weight ratios to bone, and provides for better osteointegration (the ability of bone to grow into the implant), resulting in better performance and quicker recovery.

Lattice, mesh, and cellular structures are a simplified version of topology optimisation, which consists of transforming a solid into a 'truss' structure. Most engineers intuitively understand that trusses are useful for increasing the rigidity or strength of a beam while reducing its weight (Fig. 9.16).

There are four main techniques for applying lattice structures to additively manufactured parts:

Convert the entire part into a lattice:

This technique transforms the entire volume of the part into a lattice structure. It is commonly done for medical implants, and parts where the exterior surface of the part is not critical (Fig. 9.17).

Fill the inner body of the part with a lattice structure, leaving an outer shell of a specified thickness:

In general, this method requires salt-shaker holes so that the unsintered powder can be removed from inside the part. If designed correctly, the internal lattice structure also acts as the support material for heat transfer within the part (Fig. 9.18).

The part is subdivided into solid and lattice areas:

Here a conscious decision is made as to which features of the part remain solid, and which get converted to lattices. The easiest way to achieve this is usually to split the

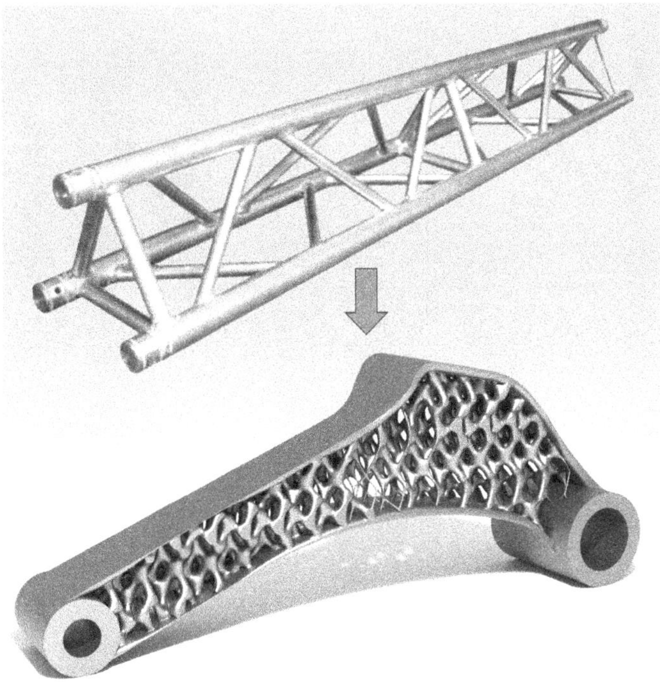

Fig. 9.16 Trusses and lattice structures. Courtesy of Autodesk

Fig. 9.17 Complete lattice structure

Fig. 9.18 Interior lattice structure

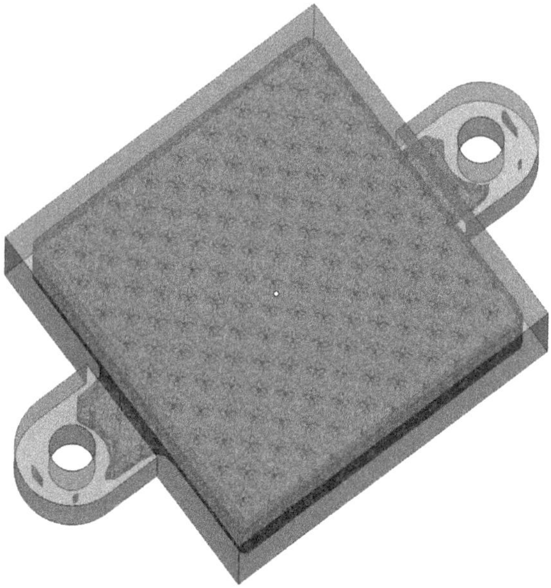

Fig. 9.19 Hybrid lattice structure

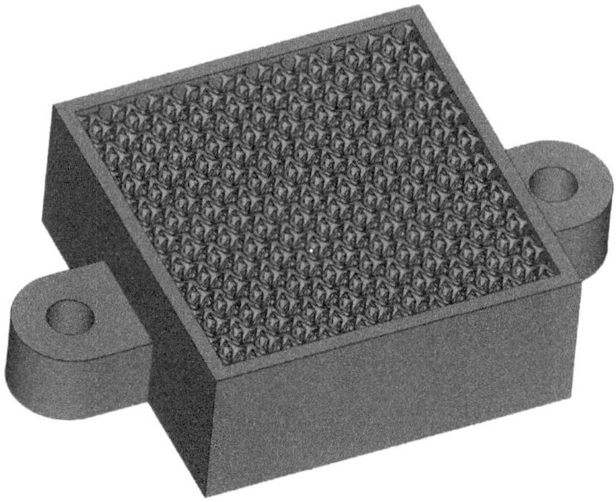

part up into its different regions in the native CAD software the part was created in, and to then import the separate parts into the lattice conversion software to convert the required parts into a lattices while leaving those that must remain solid untouched. Once this is done, a Boolean operation can be performed to join the lattice parts and solid parts to form a single part ready for AM (Fig. 9.19).

Variable lattice structure based on FEA results:

This uses any of the above techniques but, instead of a constant cell-sized lattice, uses a lattice structure where the cell size and spacing varies based on a finite element

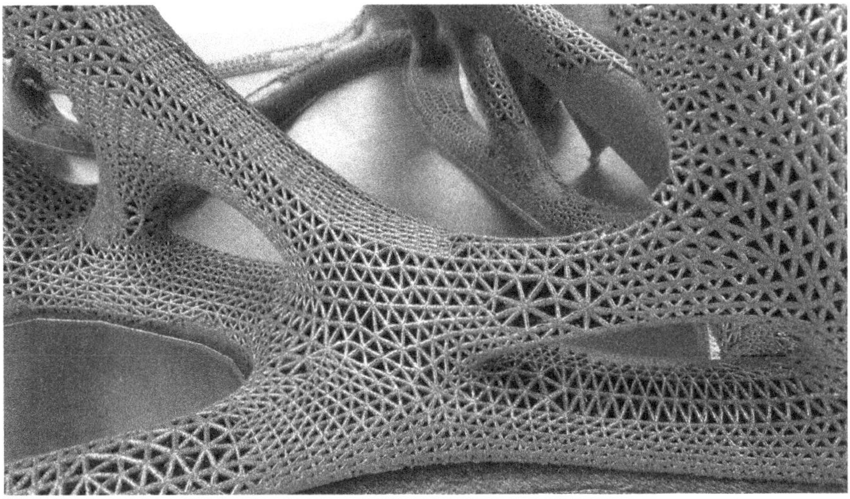

Fig. 9.20 Finite element analysis driven dynamic lattice structure, created in Optistruct by Altair. Courtesy of Altair Engineering Inc. (Nasdaq: ALTR)

analysis to produce a dynamically-sized lattice structure. The more highly stressed areas of the part wither use thicker lattice members, or a denser spacing of the lattices (Fig. 9.20).

9.13.1 Lattice Structure Strut Diameters

The strut diameters used in lattice structures must be of a diameter such that they can both be manufactured as well as provide the required mechanical properties to the part. Theoretical minimum strut diameter for metal AM is around 0.15 mm. Common sense, however, tells us that a 0.15 mm strut will have relatively little mechanical strength or resistance to fatigue. A more sensible minimum strut diameter to use is, therefore, between 0.5 and 1 mm.

When designing lattices, it is important to uses lattices that are self-supporting and can be printed without requiring support material. It is possible to have horizontal struts, but they must be short enough to have a surface area below that which requires support material.

In the lattice cell design shown below, if the forces are applied in the directions shown, then design B will resist the force better than design A, but may not be printable because of the horizontal strut. If, however, the cell is printed after being rotated by 90° as shown in C, it will still resist the forces and it will print better. The slight anisotropy in the vertical direction may, however, be a problem if the struts are very thin (Fig. 9.21).

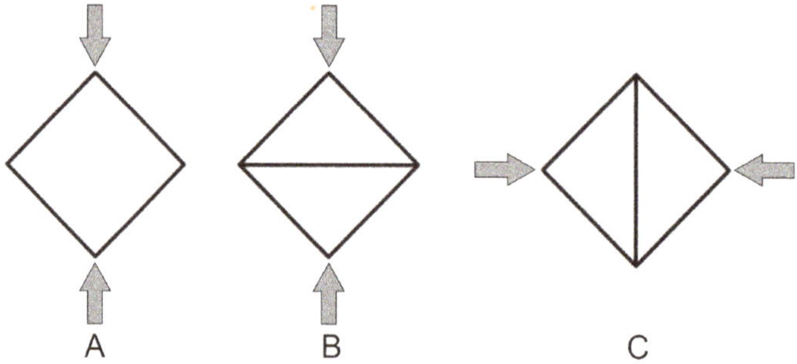

Fig. 9.21 Lattice cell A is weak against the forces, Lattice B is stronger but hard to print because of the horizontal strut. Lattice C both resists the forces and is easier to print

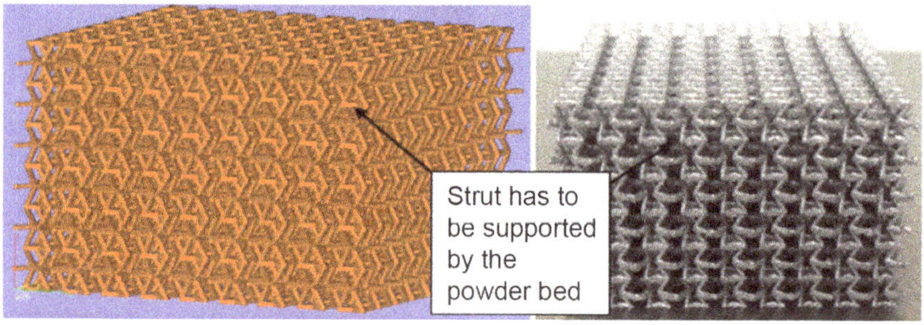

Fig. 9.22 Sagging struts caused by overly long horizontal lattice members. Courtesy of Renishaw

The design below shows an example where, depending on the cell size, the horizontal struts may sag if the horizontal span is too long to be self-supporting (Fig. 9.22).

9.14 Overhangs and Support Material

Though it may sound like more repetition, it is important to emphasize that supports are absolutely critical in the production of metal AM parts. Supports should not only always be considered during part design, but can become one of the factors that influence AM part design. The angle and surface area of any overhanging feature of the part determines whether or not the part will require support material.

In metal AM, support structures have several functions:

- support the part in case of overhangs.
- strengthen and fix the part to the building platform.
- conduct excess heat away.
- prevent warping or complete build failure.
- Prevent the melt-pool from sinking down into lose powder.
- Resist the mechanical force of the powder spreading mechanism on the part.

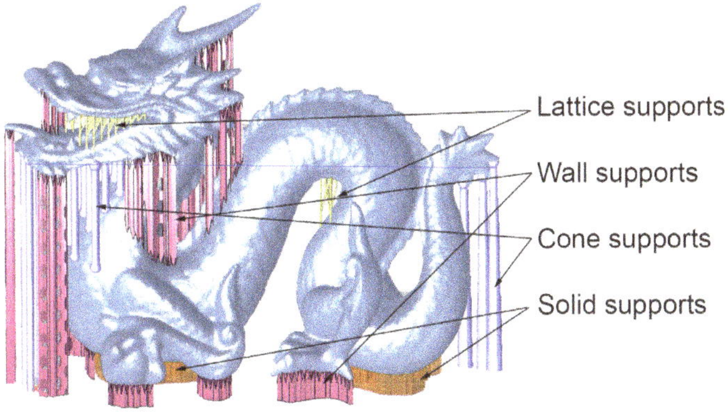

Fig. 9.23 Examples of different support types. Note that these are just used to show examples of different types of support material, and may not be ideal for this particular part

It is almost always a trade-off to orient the part for minimum build time, easy to remove supports (particularly from inside the part), surface quality, and part warpage. Some aspects will improve, while others deteriorate, depending on the support material being used.

Most metal AM pre-processing software allows the selection of a number of different support types, each of which has different heat transfer and mechanical strength characteristics. Some of the support types offered by most AM preparation software include solid, walls, trees, cones, lattices, blocks, points, lines, webs, and gussets (Fig. 9.23).

Which type of support to use very much depends on the part geometry, and how much residual stress it will contain (see the chapter on residual stress), and how hard the support material will be to remove. We are now beginning to see software that will automatically choose the most appropriate supports depending on a number of factors including the stress on the part, surface finish, and the difficulty to remove supports. This software, however, is still in its infancy but, we hope, will become a standard feature of any AM software in the future.

The best advice to understand the effect of different types of support offered by your system is to design a part made up of a series of bridges, under which each bridge is printed with a different type of support. The effect of each type of support can then be observed, both in its impact on surface finish and on removal difficulty.

9.14.1 Printing Parts with Large Horizontal Surfaces

Parts with large horizontal areas of material will, in general, require much stronger supports than the rest of the part. This is because the sudden change in cross section to a large molten sheet of material will cause substantial residual stress and will, in all likelihood, cause cracks in the part if the support is not very strong and dense. In such situations, one must sometimes use 'solid' supports, where the support is, effectively, solid material but of a slightly lower density than the main part. If at all possible, parts should try to be oriented so as to avoid large horizontal flat areas, or sudden changes in surface area to be melted, to reduce the need for solid supports, and to minimise the risk of cracks in the part.

The part below, for example, was printed in a horizontal orientation with block supports, and cracks can be seen in the bottom surface. However, as the horizontal orientation is the one with the lowest total height, the time to print three of the parts was 22 h (Fig. 9.24).

In contrast, the same part, when printed at an angle of 45 degrees, has no cracks on the bottom surface, but the three parts take 54 h to print (Fig. 9.25).

This, again, demonstrates the typical type of compromises that must be made in additive manufacturing. The horizontal orientation produces a lower build and therefore a shorter build time, but the angled orientation produces a better part but longer build time.

Fig. 9.24 Stress cracks on the bottom of a part printed in a horizontal orientation

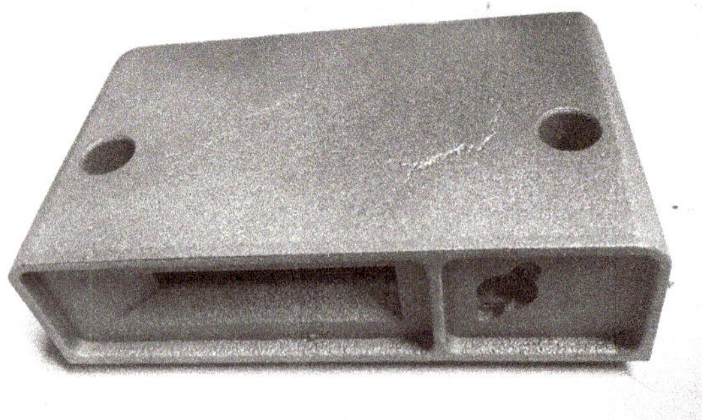

Fig. 9.25 Stress-crack free part printed at 45° orientation

9.14.2 Angle for Support Material

A general rule of thumb for angles that do not require support material are angles greater than 45° from horizontal. This does, however, vary from material to material and machine to machine. Specific angles relevant to particular materials are given in the table on "Feature type: overhang angle" later in this chapter.

Be aware, also, that some manufacturers specify angles from horizontal, while others specify them from the vertical.

Remember also, that these angles represent the minimum angle at which the part can be produced without supports. In general, using angles that are steeper than those minimum will yield parts with better surface qualities. This is because 'down-facing' surfaces always have worse surface finish than 'upwards-facing' surfaces. And the more downward the face angle is (the closer to horizontal), the worse the surface finish will be.

9.14.3 Unsupported Angles, Overhangs, and Bridges

The area melted at the focal point of the energy beam cools very quickly and the stress generated tries to curl the material upwards. Supports act as an anchor to the build plate to avoid such upward curl. The warping/curled area can also cause build failures because it prevents the spreading of a new layer of powder.

Angles

Poor surface roughness is the result of building directly on loose powder instead of using the support structure as a building scaffold. It occurs because the laser penetrates the powder bed and agglomerates loose powder around the focal point instead of dissipating the heat through the support structure. At a certain point, the unsupported angle becomes such that the part either has an unacceptable surface quality, or crashes the recoater mechanism (Fig. 9.26).

Overhangs

Overhangs differ from self-supporting angles in that they are abrupt changes in a part's geometry, such as a small feature that protrudes horizontally at 90°. Powder bed fusion is fairly limited in its support of overhangs when compared to other 3D printing technologies. In general, any design with an overhang greater than 0.5 mm (0.020 in.) will require additional support to prevent damage to the part.

As an overhang extends past about 0.5 mm, the surface quality either becomes unacceptable, or the upwards curl can become such that it causes the recoater mechanism to crash (Fig. 9.27).

Bridges

A bridge is any flat down-facing surface that is supported by 2 or more features. The minimum allowable unsupported distance for the powder bed fusion process is around 2 mm (0.080 in.).

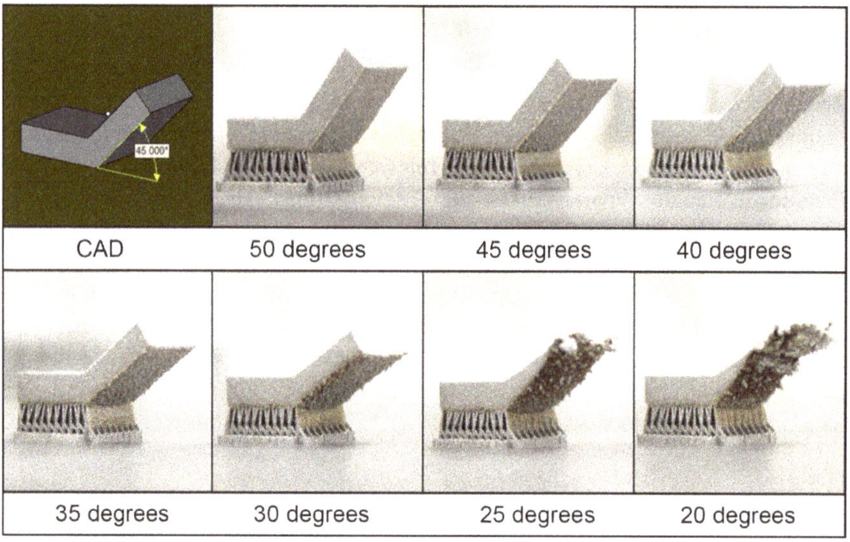

Fig. 9.26 Effect of unsupported angles. Courtesy of David Bentley, Protolabs, A Deep Dive into Metal 3D Printing

Fig. 9.27 Effect of unsupported overhangs. Courtesy of David Bentley, Protolabs, A Deep Dive into Metal 3D Printing

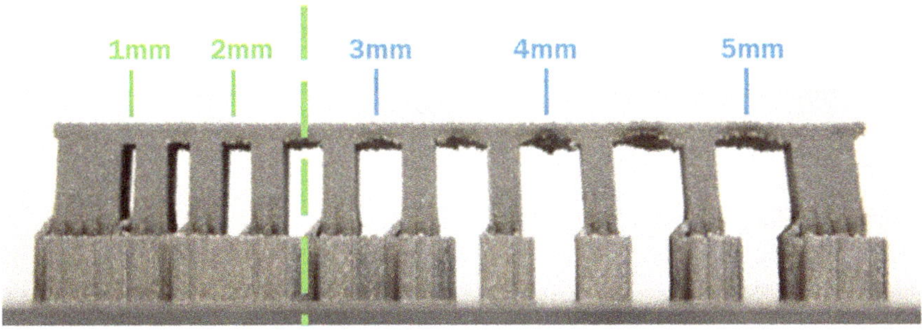

Fig. 9.28 Effect of unsupported bridges. Courtesy of David Bentley, Protolabs, A Deep Dive into Metal 3D Printing

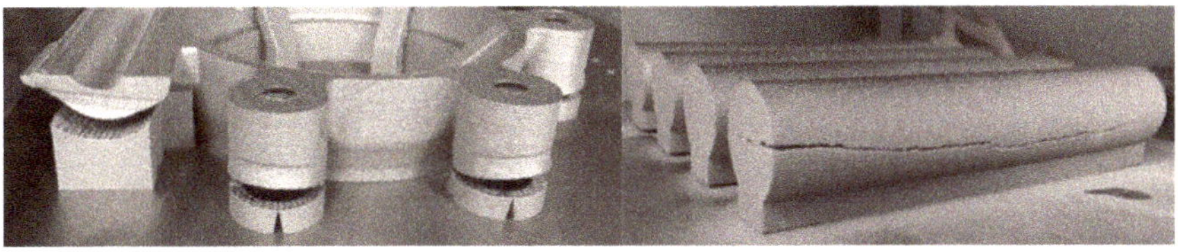

Fig. 9.29 The effect of residual stress on metal AM parts can be strong enough to detach the part from its supports, or to split the part

In the picture below, you will see how the bridge pulls in the supporting structures as the unsupported distance increases. Parts that exceed this recommended limit will have poor quality on the downward facing surfaces and may not be structurally sound. They can also cause the recoater mechanism to crash (Fig. 9.28).

9.15 Residual Stress

One of the most challenging aspects of producing good metal AM parts is residual stress. Like any welding process, metal AM induces a substantial amount of stress on the parts. This is one of the reasons why support material is often needed on metal parts. This residual stress and stress concentrations must be relieved through heat-treatment before the parts are removed from the build plate (Fig. 9.29).

Residual stress can, in some cases, be so large that it causes the entire build plate to bend, or the part to detach from the build plate, or crack the part itself.

Residual stresses are stresses that remain in a solid material after whatever caused of the stresses has been removed. An important point to remember about residual stresses is that they are not necessarily bad. They can be desirable or undesirable. Shot peening, for example, imparts beneficial compressive residual stresses into metal components. Unintended, or uncontrolled, residual stress in a designed structure, however, may cause it to fail prematurely.

Residual stress can occur from a variety of mechanisms including:

- temperature gradients existing from the surface to the centre of AM part during cooling (particularly in large masses of material) where the inside of the part cools slower than the outside of the part.
- inelastic (plastic) deformations.
- structural changes (phase transformation).
- Heat from the laser may cause localized expansion which, in AM, is taken up by either the molten metal or sections of the part that have already solidified. When the finished part cools, some areas cool and contract more than others, leaving residual stresses.

The very best solution to combatting residual stress is to try and eliminate as much of the residual stress out of the part as possible through its design.

9.15.1 Designing to Reduce Residual Stress

There are a number of relatively simple design techniques that can be employed to minimize residual stress. These include:

- Get rid of areas of uneven thickness. Large masses of material are the single biggest, but easily avoidable, source of residual stress.
- Try to avoid large changes in cross-section. This may, sometimes, mean having to print your component at an orientation other than horizontal.
- Pre-heat the build plate.
- Heat the build chamber.

In addition, many of the design rules for conventional casting apply equally to metal AM.

If large masses of material are completely unavoidable (which is rare), use different laser hatch parameter settings to minimize the build-up of residual stress.

- Smaller chess-board hatch patterns will, for example, create less residual stress than bigger ones, or than large scan areas. But they will slow down the build process a bit.
- Rotate each hatch scan, usually by 67°, for each layer.

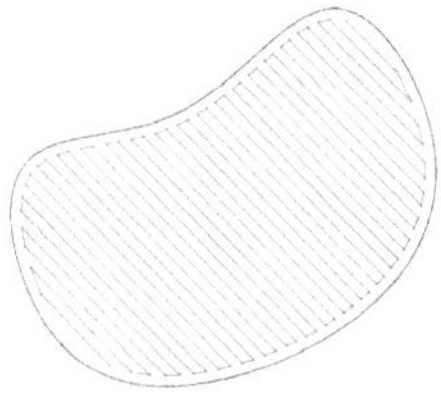	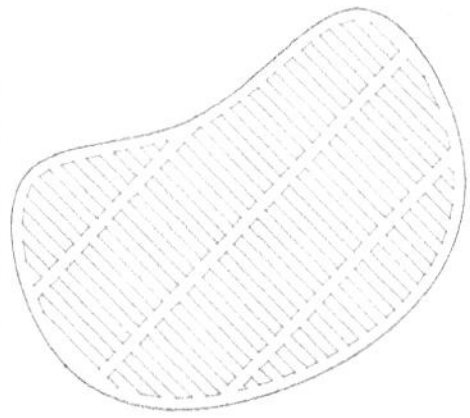	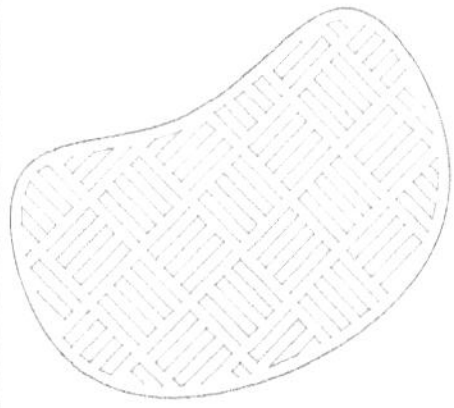
Meander hatch pattern High build rate Higher residual stress Suitable for small/thin parts	**Stripe hatch pattern** Medium build rate Medium residual stress Suitable for large parts	**Chessboard hatch pattern** Slow build rate Lower residual stress Suitable for large parts

Ultimately, some residual stress is unavoidable. The real question, however, is whether it will affect the function of the part. If it must be reduced, then there are various forms of heat treatment that can be used to reduce the residual stress in the part. This is discussed further in the chapter on post-processing. A well designed for AM part, however, can often require minimal, or no, heat treatment compared to a part that has not been designed for AM.

9.15.2 Designing to Minimize Residual Stress Example

A 2018 LinkedIn post by Alexander Lu, a well-known metal AM trouble-shooter from Temasek Polytechnic in Singapore, demonstrated the problem of residual stress with the failed print of a topology optimized bracket in which the parts kept either detaching from the build plate, or ripping away from the support material. This was caused by the considerable residual stress within the parts that was so severe that it ripped the parts off of the build plate or supports (Fig. 9.30).

The original bracket was designed as a topology optimized part made up of two components: a small clamp, and the main bracket. The fact that a part has been run through topology optimization software does not, however, automatically make it suitable for metal AM. In this case, the main part contained both large masses of material and many uneven thicknesses, that caused substantial stress, as well as several sharp internal corners, that cause stress concentrations (or stress raisers) that may have caused premature failure of the part (Fig. 9.31).

One of the most fundamental design guidelines that applies to many manufacturing methods including injection molding, casting, and additive manufacturing is to use an even wall-thickness wherever possible. In additive manufacturing this simple technique can have an even greater impact than on other manufacturing methods.

On the smaller clamp part of the bracket this simple technique of removing unnecessary material, and using ribs to maintain mechanical integrity, was used to achieve an even wall thickness of 3 mm for the main body of the clamp, a common wall thick-

Fig. 9.30 Examples of parts detaching from the build plate or support material because of residual stress. Courtesy of Alexander Liu

ness used in light to medium-duty casting applications. The ribs were 2 mm wide so as not to create an uneven thickness where the rib joined the cylinder. In addition, the four bolt hole lugs were shelled out and ribbed to avoid them becoming an area of residual stress. Depending on the particular application context for this bracket, the nominal 3 mm thickness could, of course, be either increased or decreased as required (Fig. 9.32).

For the main part of the bracket, the topology optimized version was used as the design source for a parametrically designed CAD version. The clamp ends of the brackets were redesigned very much in the same way as the above described small clamp. The thick centre member was split into 3 even wall-thickness lofted members, and the centremost of those members was designed as a U section for extra rigidity. Again, all thicknesses were kept to a thickness of 3 mm. In addition, every single

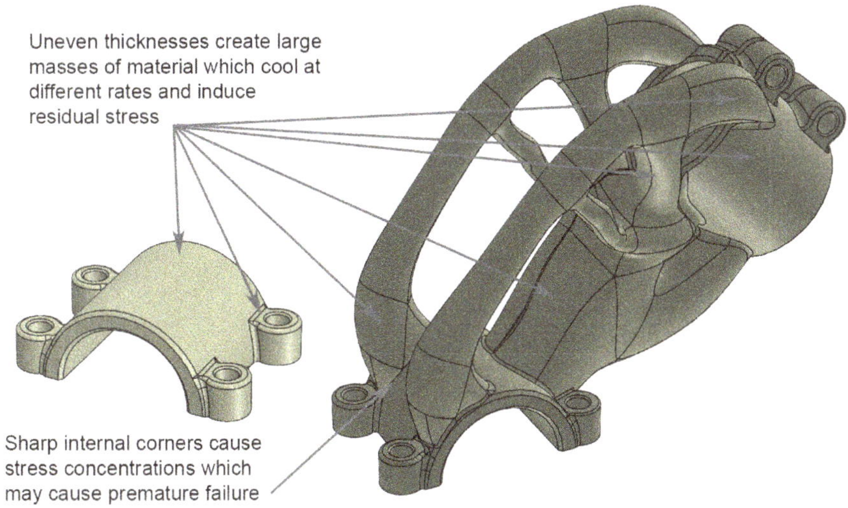

Fig. 9.31 Some of the problems identified in the original bracket and clamp designs

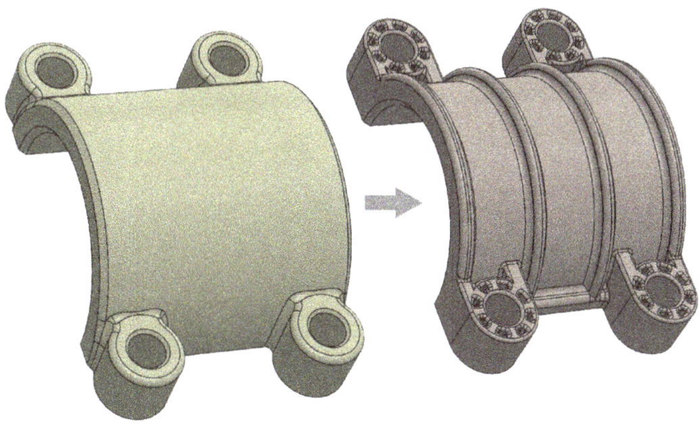

Fig. 9.32 Redesign of small clamp. The new design has an even wall thickness and no areas where stress can build up. It also weighs 25% less than the original

internal corner was filleted to avoid the risk of stress concentrations at the joints (Fig. 9.33).

The new design was successfully printed in two different orientations without any lifting off the build plate or breaking away from the support material whatsoever. The smaller clamp was printed in a horizontal and in a vertical orientation, while the larger bracket was printed in one orientation with one of the clamps in the horizontal position, and in the second orientation with both clamps at an angle to achieve a lower print height (Fig. 9.34).

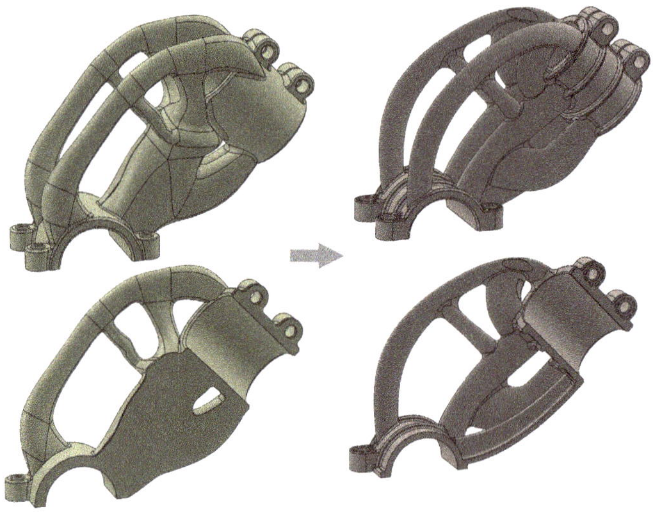

Fig. 9.33 Redesign of main bracket from originally topology optimized bracket. New bracket has relatively even wall thicknesses everywhere. The new design weighs 47% of the original bracket

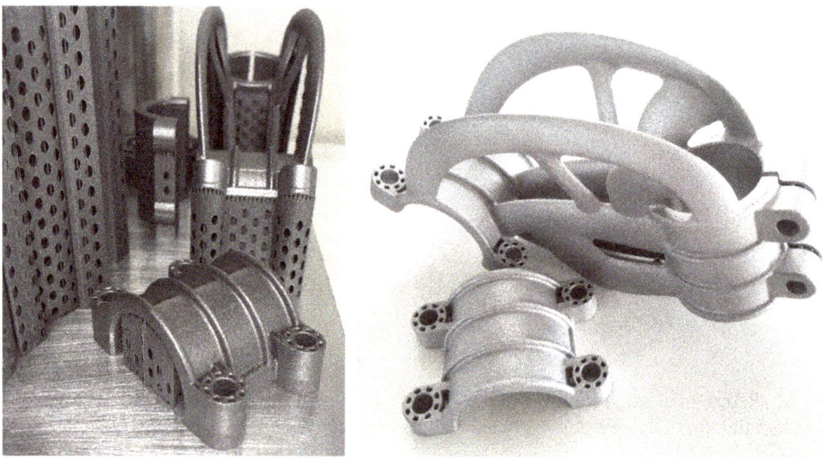

Fig. 9.34 Printed parts still attached to build platform with supports, and parts after post-processing and shot-peening

When designing parts that will be manufactured through AM, techniques for avoiding residual stress must be constantly present in the mind of the engineer. But, applying these often simple techniques, such as keeping to an even wall-thickness and avoiding large masses of material, can have a dramatic impact on the ability to successfully print the parts.

Reducing the residual stress that can build up in a part during manufacture can also greatly reduce the need for heat treatment after the build has finished. Not only that, but avoiding large masses of material, as described in a previous article in Metal AM Magazine, also has a substantial impact on reducing the time it takes to manufacture the part, and therefore its cost.

The example above clearly demonstrates that it is not practical to take conventionally designed parts (even if they do employ modern design techniques such as topology optimization) and expect them to perform well in an AM manufacturing environment. Design for AM is not a choice. It is an absolute necessity to make best use of the true power of additive manufacturing.

9.16 Stress Concentrations

A stress concentration, or stress raiser, is a location in a part where stress is concentrated. These stresses occur both within the AM fabrication process, and in the heat treatment of AM parts.

With metal AM, this is a design opportunity where a well-designed part can minimize the areas of stress concentration. Fatigue cracks almost always start at areas of stress concentration, so removing the areas in which such defects can occur can minimize such defects and greatly increases the fatigue strength of the part (Fig. 9.35).

9.16.1 Designing to Reduce Stress Concentrations

The best way of minimizing the amount of heat treatment required is to design your parts to have as little stress induced or concentrated in them as possible. Simple strategies, like filleting all sharp corners (reduces stress concentrations), even wall thicknesses, and avoiding large masses of material (reduces residual stress), can help a lot.

In the simple part below, for example, the sharp internal corner has a good chance of causing a stress crack. In addition, the sharp corner has a larger mass of material than the horizontal and vertical walls and will, therefore, contain some residual stress which could cause the wall to distort. In contrast the filleted corner has eliminated the possibility of a stress crack, and it's even wall thickness has minimised the potential of residual stress (Fig. 9.36).

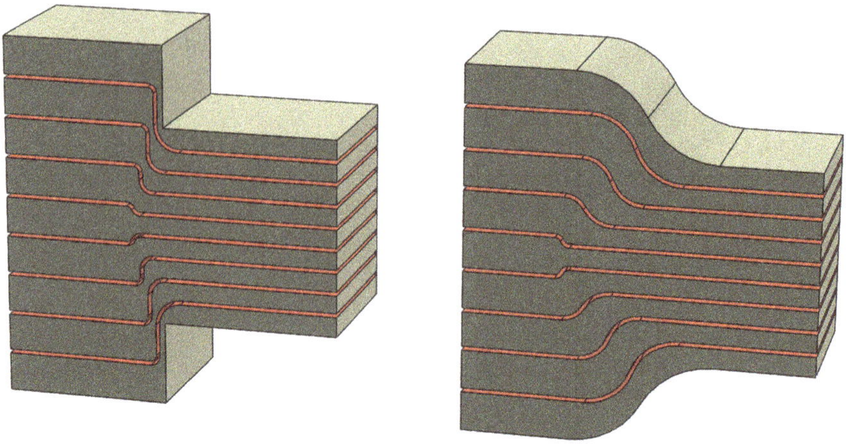

Fig. 9.35 Stress concentrations are the areas where cracks will most naturally form. Eliminating such stress concentrations can be critical to part quality

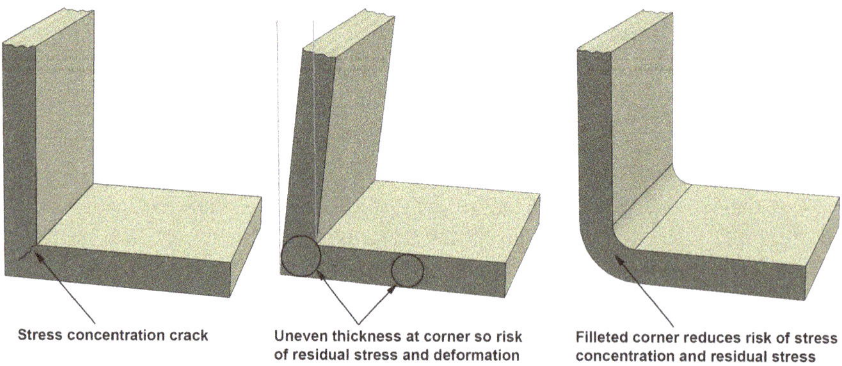

Fig. 9.36 Example of simple filleting to eliminate both stress concentrations and residual stress

9.17 Horizontal Holes

In metal AM, horizontal holes (or holes angled below the minimum support angle) over a certain diameter will require support material inside the hole. Though this is not necessarily a problem, it should be remembered that it is always harder to remove support material from inside the part than from outside the part. For long holes or pipes that are not perfectly straight, in particular, the support can be hard to remove from inside the pipe. As a general guideline, holes below a diameter of 8 mm (0.314 in.) can be printed without supports (Fig. 9.37).

If larger holes are required, the most common technique is to change the hole from a circular to a shape that can be printed without the need for support material. These shapes commonly include ellipses, teardrops, and diamonds (Fig. 9.38).

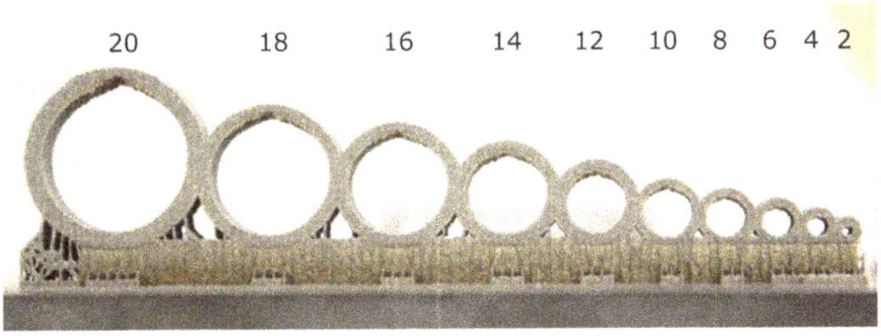

Fig. 9.37 Horizontal holes printed without support material. Courtesy of David Bentley, Protolabs, A Deep Dive into Metal 3D Printing

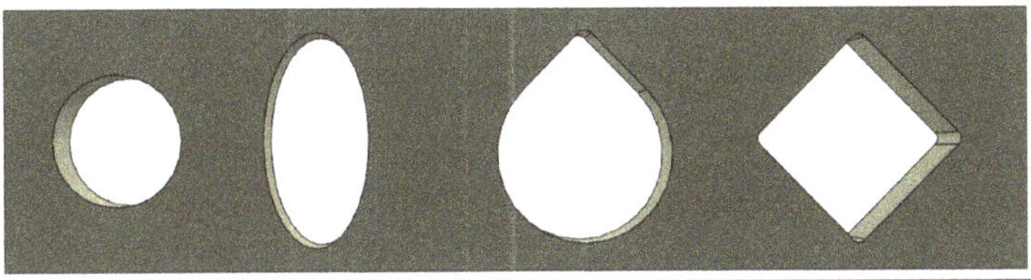

Round holes can, generally, be built without support up to a diameter of around 8mm. Holes larger than this will require supports. Note that this diameter varies based on the machine and material used.	Elliptical holes, when the height of the ellipse is twice the width, can be printed to about 25mm tall, depending on the system being used.	Teardrop shaped holes can be printed to almost any diameter providing the top angle is no less than the minimum support angle. It is good practice to fillet the top of the teardrop to avoid a stress concentration.	Diamond shaped holes can be printed to almost any size. It is good practice to fillet the corners of the hole to avoid stress concentrations in the corners.

Fig. 9.38 Hole shapes that can be printed without the need for support material

9.18 Setting up a Metal AM Print Job

The process of setting up a metal AM print job can be complex. Though it is rare for engineers and designers to be directly involved with the printing of their parts, they can greatly benefit from hands-on operation of the machines, because this will give them a much better idea of the complexities involved. If the designers of the parts to be printed are not directly involved with the printing, it is critical that they have a good dialog with the machine operators to ensure that the resulting parts meet their expectations.

On most metal AM systems, the print-job occurs on a build plate. This plate can be either square or round, and is bolted onto the build chamber piston which will move it down by the thickness of a single layer after each melting operation has completed. The powder is spread over the build plate in a thin layer by the 'recoater' system which consists either of a soft (silicone or carbon fiber brushes) or hard (steel or ceramic) wiper blade, or a hard roller.

How one positions the parts on the build plate can have a major impact on the success and quality of the printed parts.

As the recoater spreads a new layer of powder, the sections of the part that have already been printed cannot move, so if there has been any distortion (even minute distortion) on the part, it can exert some mechanical force on the recoater. This force can be enough to bend or break the parts if the features are delicate. If the part distorts but doesn't bend or break, it can be enough to crash the recoater, causing a build failure.

By correctly positioning your parts on the build platform, you can minimize the amount of force the parts exerts on the recoater, and therefore reduce or even eliminate failed builds related to recoater crashes.

9.18.1 General Part Positioning Guidelines

If a part is largely rectangular, or contains long flat walls, try not to position your parts with long walls parallel to the recoater, as this will cause the recoater to suddenly meet a large obstacle. Instead, rotate them around the vertical Z axis to try and minimize the recoater force at any one point.

If you put your part parallel to the recoater and there is any distortion in the part, the recoater may not be able to pass over the distortion and will crash the build. Rotate your part around the Z axis by between 5° and 45° so the recoater doesn't suddenly meet a long flat wall. This will greatly reduce the risk of a crash and can improve the quality of delicate features such as thin walls (Fig. 9.39).

Try to avoid putting parts right behind each other. If a part distorts and makes contact with the recoater, the build might continue even though the recoater, or the part, may be damaged. This is especially the case with silicone or carbon fiber brush blades.

The effect of this may be a deterioration in the spreading of the powder right behind the collision area. So, if possible, when nesting your parts on the build plate, try to add some space behind your parts along the recoating axis (Figs. 9.40 and 9.41).

Try to avoid having the recoater make simultaneous contact with several parts at once. Just shifting your parts on the build platform by a few millimetres is often enough to minimize the risk of distorted geometries crashing the recoater system (Fig. 9.42).

Position the highest parts/areas closest to the recoater. The reason for this is simple pragmatism. On some printers, you cannot put enough powder in the machine to print

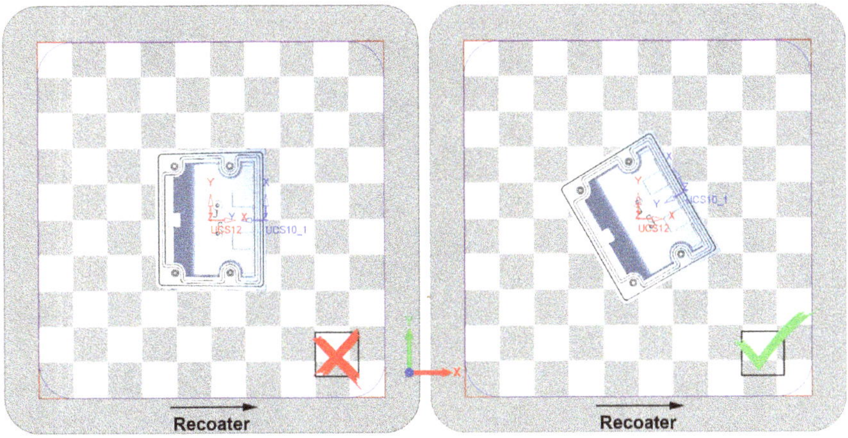

Fig. 9.39 Avoid parts parallel to the recoater blade

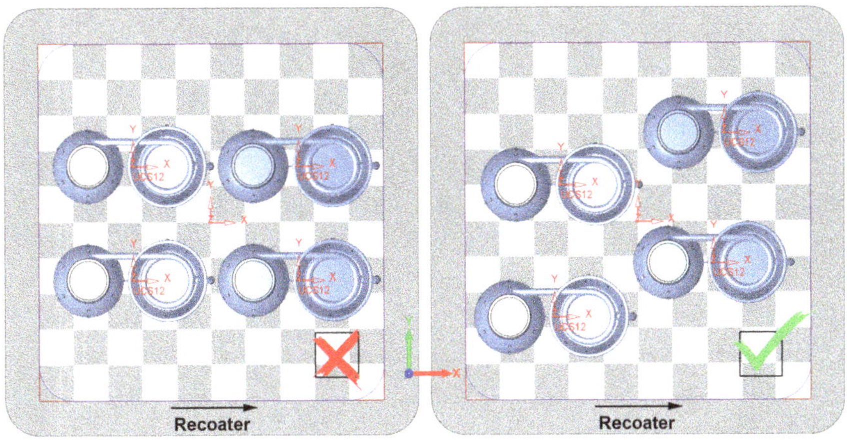

Fig. 9.40 Avoid parts lined up directly behind each other

very high parts or a full build chamber (incredible but true!). To print such tall parts, the machine may need to be paused while powder is added part-way through the build. However, some of these printers may also allow you to reduce the amount of powder deposited in each layer. In this case, one would use the 'normal' amount of powder while the small parts are being printed and, once they are finished, the amount of powder per layer is reduced, hopefully to the point where there is enough powder to print the entire build without needing to pause and refill the machine which is undesirable due to parts shrinking during the cooling time needed to refill the system (Fig. 9.43).

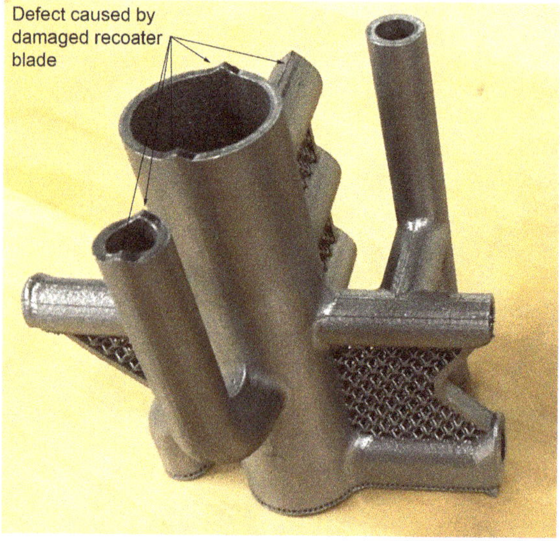

Fig. 9.41 Example of part in which the recoater blade was damaged and cause d a defect in the part

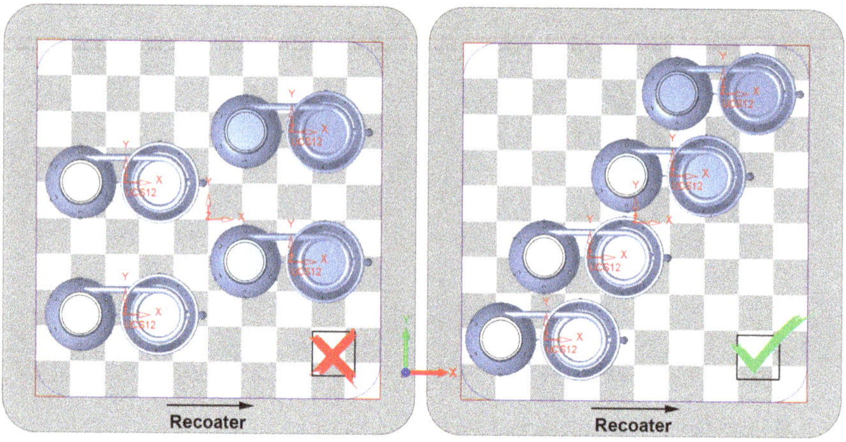

Fig. 9.42 Avoid multiple parts hitting the recoater blade at the same time

Something to be aware of with the above guidelines is that the print sizes advertised by the system manufacturers may sometimes be misleading. If a manufacturer quotes, for example, a 250 mm × 250 mm print size, but requires that parts be positioned at an angle, then a 250 mm × 250 mm square part would not fit in the advertised build volume. It may 'theoretically' fit, but would not be printable because its orientation may cause the machine to crash.

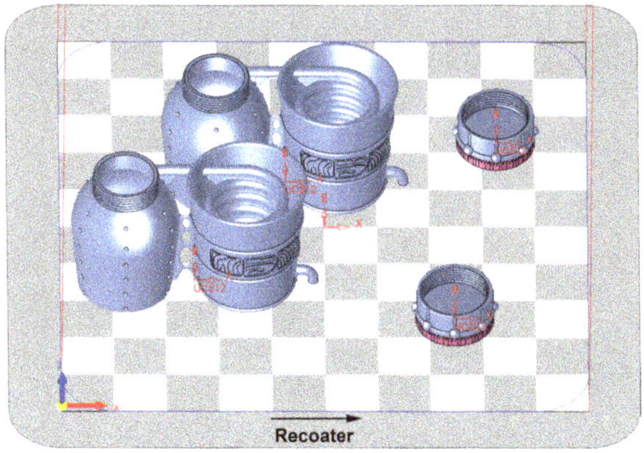

Fig. 9.43 Position the tallest part closest to the recoater blade

Chapter 10
Metal AM Guidelines

10.1 Design for Laser Powder Bed Fusion

The design guidelines below apply to laser powder bed fusion metal processes. The guidelines will vary from machine manufacturer and model to machine manufacturer and model so, if in doubt it is recommended to print a test piece to verify each set of design parameters.

10.1.1 Feature Type: Wall Thickness

Minimum wall thickness (t)	Recommended minimum wall thickness (t)	
0.3 mm (0.016 in.)	1 mm (0.039 in.)	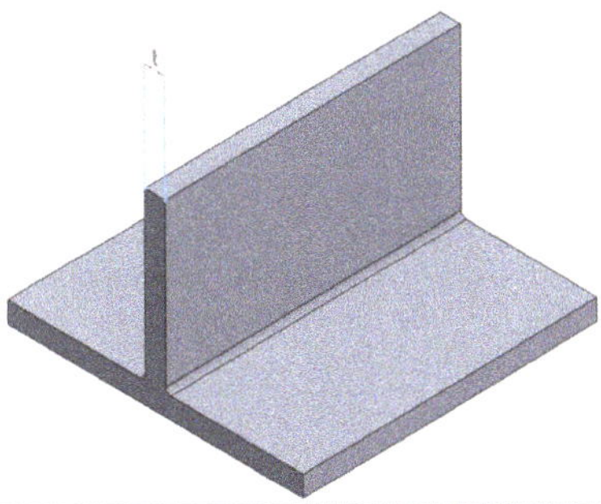

Comments

Problems may occur with extended lengths of unsupported walls (i.e., no ribs or intersecting walls). Without adequate reinforcement, large surface area thin walls are likely to distort. In this case, avoid using the minimum wall thickness, or reinforce the wall with ribs, gussets, or use extra support material to prevent it from distorting.

Always fillet the corners where the walls meet another surface. A good rule of thumb is to make the fillet ¼ of the thickness of the walls.

O. Diegel et al., *A Practical Guide to Design for Additive Manufacturing*, Springer Series in Advanced Manufacturing,
https://doi.org/10.1007/978-981-13-8281-9_10

10.1.2 Feature Type: Overhang Angle

Maximum overhang angle (a)	
DMLS stainless steel	60°
DMLS Inconel	45°
DMLS titanium	60°
DMLS aluminium	45°
DMLS cobalt chrome	60°

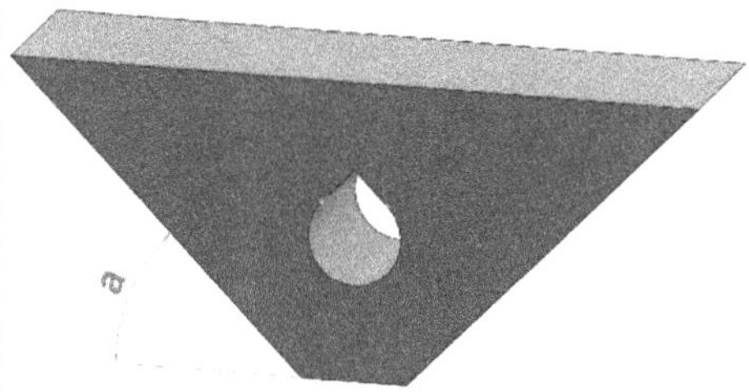

Comments

Overhang angles less than the numbers shown above (measured from horizontal) will require support material, which may be added automatically by the system software. Excessive supports that need to be removed manually will increase post-processing time.

Beware that some manufacturers measure support angles from the horizontal, while others measure it from the vertical.

Feature shapes (e.g., cooling channel profiles) can often be modified to minimize support requirements, and horizontal holes less than 8 mm (0.236 in.) can be built without supports. Se the chapter of this book on the design guidelines for horizontal holes

10.1.3 Feature Type: Clearance Between Moving Parts

Minimum clearance	
Horizontal (h)	Vertical (v)
0.2 mm (0.079 in.)	Adequate access to facilitate the removal of support material

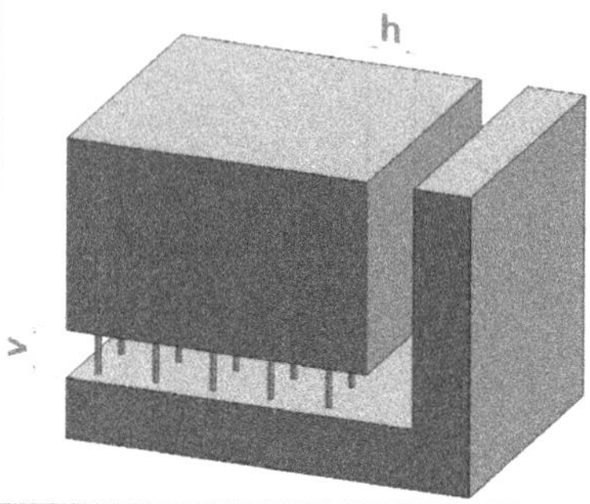

Comments

In general, with metal AM, all moving parts will need to be welded to the build platform, or connected to each other, so that they are now swept away by the recoater system. They only become moving parts once they have been cut off of the platform or the joining links have been cut.

Large areas of close proximity will make the removal of supports more difficult. Clearance between parts built separately and assembled later must be at least equal to the general build tolerance of the system.

10.1.4 Feature Type: Vertical Slots and Circular Holes

Minimum width for slot (w)	Minimum diameter for circular hole (d)	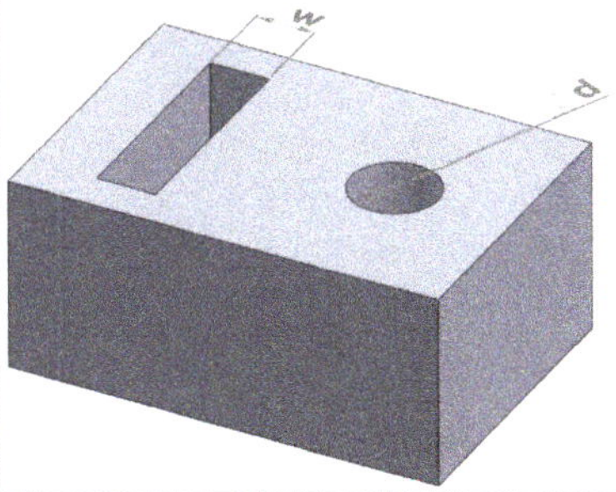
0.5 mm (0.02 in.)	0.5 mm (0.02 in.)	

Comments

As the thickness of the part increases powder inside the slots or holes may become hard, or impossible, to get out.

Values for horizontal features are not available as they are too dependent on each specific machine.

If possible fillet all sharp internal corners to avoid stress concentrations.

10.1.5 Feature Type: Vertical Bosses and Circular Pins

Minimum width for boss (w)	Minimum diameter for circular pin (d)	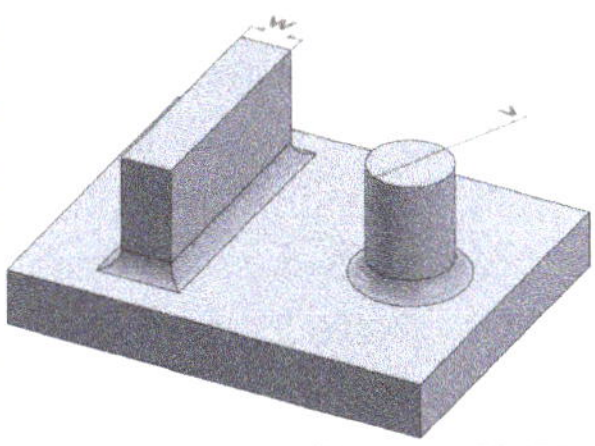
0.5 mm (0.02 in.)	0.5 mm (0.02 in.)	

Comments

It is good practice to fillet the bottom of all pins and bosses. A general guidelines is that the radius is ¼ of the thickness.

Values for horizontal features are not available as they are too dependent on each specific machine.

10.1.6 Feature Type: Built-In External Screw Threads

Threads should always be built vertically, if at all possible.	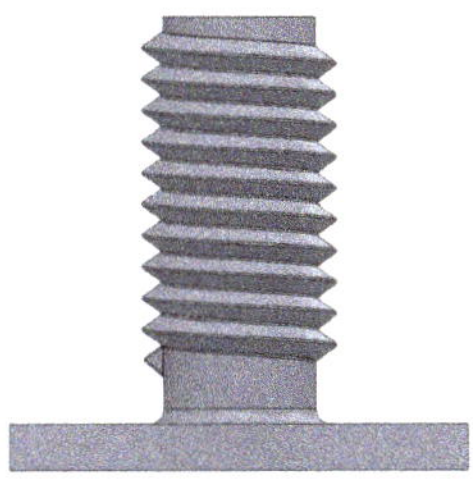

Comments

Though threads down to about M4 can, theoretically be printed, their surface roughness means they will need to have a tapping die run over them to clean up the thread.

Tapping is recommended for all threads and sufficient space must be left around the post to allow access for the tapping die.

Fillet the bottom of the boss where it meets the wall to avoid stress concentrations. A good rule of thumb is to make the fillet ¼ of the thickness of the walls.

10.2 Design for Electron Beam Melting

Electron beam melting is a powder bed fusion process that uses an electron beam as the energy source to melt each layer of powder. The electron beam is controlled by electromagnets that move it over the powder in a controlled manner to draw the slices of the part to be produced.

Typically, an EBM process will include the following steps 5:

1. **Inert the build chamber**: All air is evacuated from the build chamber with a vacuum pump. This can take from about 50 to 70 min. All door seals should always be thoroughly cleaned in order to ensure no leakage can take place when the door is closed. The vacuum system provides a base pressure of 1×10^{-5} mbar or better throughout the entire build cycle. During the actual melting process a partial pressure of Helium is introduced to 2×10^{-3} mbar. As with all metal AM processes, a clean and controlled build environment is important to maintain the chemical specification of the metal material.
2. **Ramp up the electron gun**: The electron gun filament is heated up by current. All electron gun components should be free of any metal powder particles in order to avoid arcing. This part of the process takes about 10 min.
3. **Calibration**: The electron beam spot is adjusted to be of the correct size. This can take approximately 5 min.
4. **Preheating of the build plate**: A temperature of about 750 °C (1380 °F) needs to be reached before the print job starts. This can take about 40–50 min.
5. **Preheating the layer**: For each layer in the build the electron beam preheats the entire powder bed to an optimal ambient temperature, specific for the material used. It does this by scanning the electron beam over the entire layer, and partially sintering the metal powder together. As a result, the parts produced with the EBM process are relatively free from residual stresses and have a microstructure free from martensitic structures.
6. **Melting**: Once the entire layer has been preheated, and partially sintered, the electron beam traces out the slices of the parts and fully melts the powder in those areas.
7. **Cool down**: When the build is finished, the build chamber needs to cool for about 12–18 h, depending on the build height. Heated titanium powder can be explosive if the build chamber is exposed to ambient air while still too hot.

8. **Removal**: After the cake has sufficiently cooled down [to below 100 °C (212 °F)] the build chamber is removed from the machine, and the part, which is suspended inside the partially sintered powder cake, is taken out. The unused powder can be sieved and topped up with new powder and used again.

10.2.1 Post Processing

After a build is complete, it is removed from the machine by the machine operator. Various steps are then required before the part is ready for use.

1. **Powder recovery/Blasting**: EBM parts are removed from the machine with a semi sintered block of caked powder around the parts. Arcam/GE provides a Powder Recovery System (PRS), which is recommended for use with their EBM machines. It is a sealed air pressure blasting chamber which enables recovery of partially sintered powder from the build.

The media used in the PRS is the same as the powder used in the build, so both the media and the partially sintered powder can be mixed in the PRS process, sieved, and returned to the machine for the next build without being contaminated. The time taken to blast partially sintered powder from a build can vary greatly depending on the complexity of the parts. In hollow cylinders of varying lengths and 5–7 mm diameter, about 75% of the partially sintered powder can be removed in the first 30 s of blasting. To remove about 98% of the powder might take another 5 min. However, the more corners or blind holes a shape has, the longer powder may take to be removed by blasting and, in some cases of complex geometry it may not be possible to remove all the powder by blasting.

Other techniques for powder removal, for example using hand tools to scrape or break the powder cake away, may be sufficient to remove the cake. This will result in powder losses as it is not possible to return powder to the machine after contamination with other tools outside the PRS. An ultrasonic tool can also be an option to remove trapped volumes of powder from internal passageways.

2. **Support removal**: Fewer supports are required for EBM due to the partially sintered powder cake surrounding the part, but the supports that are required for additional heat transfer should be possible to remove in the same way as for laser based powder bed fusion parts. The empty platforms can be reused but will need to be resurfaced mechanically e.g. by machining, or EDM wire cutting.
3. **Heat treatment**: Because there is little residual stress in EBM parts, post heat treatment is not usually necessary. This is due to the partially sintered powder around EBM parts being held in a vacuum at high temperature in the EBM process, which anneals the material as the layers build up and relieves the stresses in the part as the build progresses. Surfaces that are attached directly to the build platform by supports are stress relieved during the build, but any parts that do not have supports are susceptible to higher stresses. Heat treatment can, however, be used to improve certain material properties of EBM parts.

4. **Hot Isostatic Pressing (HIP)**: HIP is well suited to EBM because the build process takes place in vacuum, therefore any internal pores also contain a vacuum and can be more easily collapsed by the HIP process.
5. **Machining**: The main points to consider when designing an EBM part are which surfaces of the part need to be machined for accuracy. Also think about additional stock material considerations for machining (usually between 0.5 and 2 mm depending on the part), and whether mounting fixtures should be added to the part design to allow it to be clamped or secured for easier machining. As with any AM, think about whether the rougher as-built surfaces can remain in some areas? Also always think about whether it will be possible to access surfaces to be machined once the part has been built?
6. **Surface finishing**: The surface finish required on the part can affect the design. Sometimes a standard blasted or shot peened finish is suitable. If better surfaces are required, all the same techniques that are used in other metal surface finishing, some of which are described in the next chapter of this book, can also be used for EBM.

10.2.2 Design Guidelines

Arcam/GE recommend that parts in EBM builds should be oriented with as few down facing surfaces as possible and, in this respect, EBM is now different to other laser powder bed fusion processes. This can also be taken into account during the design process to try and minimize the amount of down facing surfaces. Horizontal surfaces, especially curved ones, are prone to variations in dimensional accuracy and defect formation.

Supports: The purpose for supports in EBM differs slightly from laser based powder bed fusion, mainly because of the partially sintered cake material that is built around the part. This provides some level of anchoring and thermal conductivity that the loose powder in laser based powder bed fusion does not. This also reduces residual stresses inside the part.
However, the bottom of the part does needs some supports to anchor it to the platform and on some downward facing surfaces (downskins) there is a need for additional support material to prevent distortion and to maintain dimensional accuracy. Thin overhangs and horizontal edges may also result in overheating and distortion. Support structures, similar to those used in laser based processes can be added to the part to act as a heat sinks. In EBM, these are called 'wafer supports'. Denser supports are often needed on areas of large mass, because they need more heat transfer than thin walls. Unlike laser based processes, these supports do not need to be attached to the build plate as the partially sintered powder cake allows the supports to start building later in the build so wastes less material.
Details: In general, small surface details should be oriented on upward facing surfaces and free from supports. Very short horizontal bosses or overhangs protruding <1 mm

can be built without requiring supports, but the downward faces of these will be rough. Longer horizontal protrusions will require supports in EBM or warping and over-melting can occur. As-built surface texture, post processing requirements and access to detailed surfaces for finishing are important to consider. If mechanical finishing or abrasive blasting will be carried out, this may round off fine details or edges.

Machining stock: Due to the rougher surfaces produced by EBM, 0.5–2.0 mm would be a suitable addition for machining stock, depending on the orientation and roughness of the surface. More stock material will need to be removed on rough down-faces and verticals than on smoother upwards facing surfaces before a dense flat surface is achieved, so this will affect the amount of stock needed in different areas of the part.

Hollow parts: Parts built in EBM are embedded in partially sintered cake of powder. This powder is not possible to pour out and must be blasted away. Multiple access holes will be necessary to allow the internal caked powder to be removed by blasting or hand tools. A hollow open lattice form is often more suitable for EBM than a more enclosed hollow form because of the constraints of blasting the partially sintered powder out of the internal cavity. It is possible to build 'parts within parts' using the partially sintered powder cake as support.

Stacked parts: With EBM multiple components can be stacked on top of one another to maximize the number produced in a single build. They can be attached to the part below with minimal support, as they are supported in the partially sintered cake which acts as a thermal conductor to aid heat distribution without the need for full support contact with the platform or the parts below.

The table below gives a comparison of some of the design guideline differences between electron beam and laser based powder bed fusion.

Feature type	Laser powder bed fusion	EBM
Orientation	Upskins/verticals smoother, downskins rougher	Upskins smoother, Vertical/Downskins rougher
Support angle	Bottom edge and surfaces ~45° or less from platform need support	Bottom edge and horizontals or areas vulnerable to over melting and need support
Wall thickness	Minimum 0.3–0.5 mm	Minimum 0.6–1.0 mm
Details	Small details should be oriented upwards to avoid the need for supports/rougher surface textures	
Holes/tubes	>0.5 mm ø vertical; <8 mm ø horizontal without support	>0.5–2.0 mm ø can build but powder removal may be difficult
Machining stock	0.1–0.5 mm to be added to machining surfaces	0.5–2.0 mm to be added to machining surfaces
Clearance	>0.15 mm for assemblies; >1.0 mm between parts	>1.0 mm between parts removal of powder must be considered

(continued)

(continued)

Feature type	Laser powder bed fusion	EBM
Hollowing	0.3–0.5 mm min wall thickness, must add powder release hole	0.6–1.0 mm min wall thickness removal of caked powder must be considered
Screw threads	Should be built vertically, must be tapped/machined	Must be tapped/machined

10.2.3 Feature Type: Wall Thickness

Minimum wall thickness (t)	Recommended minimum wall thickness (t)	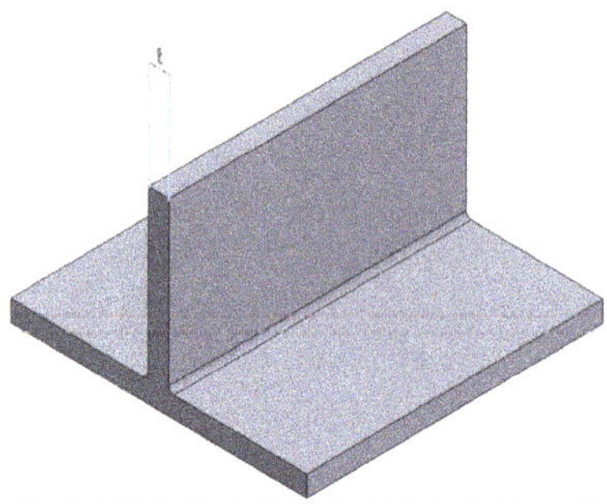
0.6 mm (0.032 in.)	1 mm (0.039 in.)	

Comments

It is possible to build vertical walls with a thickness of 0.6 mm in solid material, but this can be difficult to achieve in all orientations and with walls of large surface area. A safe recommended wall thickness is 1 mm. For short lengths, such as in lattice structures, different melt strategies can be used so the part can be as thin as 0.3 mm. However, this is not suitable for structural walls as they can suffer from delamination or layer shift.

Always fillet the corners where walls meet each other.

10.2.4 Feature Type: Vertical Slots and Circular Holes

Minimum diameter for circular horizontal hole (h)	Minimum diameter for circular vertical hole (v)	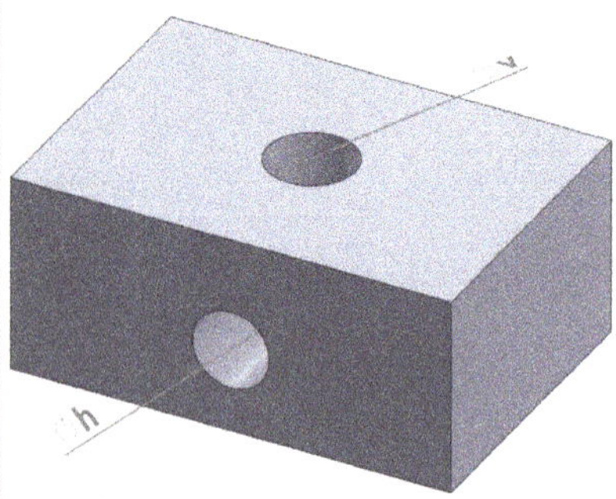
0.5 mm (0.02 in.)	1 mm (0.04 in.)	

Comments

Holes, slots or tubes built into EBM parts at any angle will be filled with partially sintered powder. This block of powder allows different diameters to be built without the need for supports but can be hard to remove unless access to the hole is easily done with blasting media or with hand tools, so this must be considered during the design stage.

A minimum diameter of 1 mm vertically or 0.5 mm horizontally is recommended to ensure that rough surfaces do not cause the holes to close up.

In walls thicker than about 2 mm, vertical holes will generally need to be no smaller than 2 mm and, for horizontal holes, no smaller than 1 mm.

10.2.5 Feature Type: Clearances to Remove Powder

Minimum clearance		
horizontal (h)	vertical (v)	
1 mm (0.04 in.)	1 mm (0.04 in.)	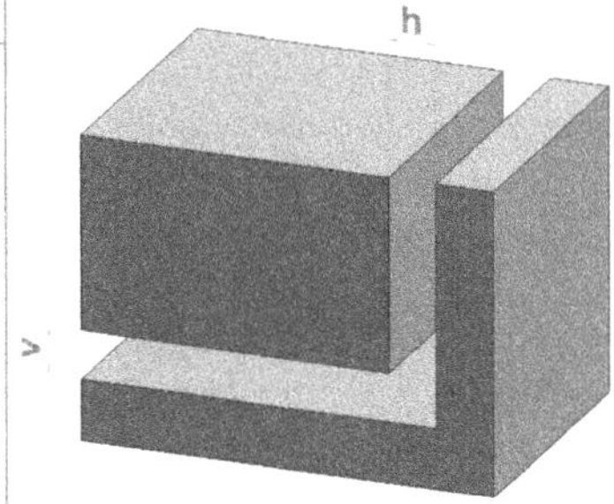

Comments

As there is a partially sintered cake of powder surrounding the parts that are built, access must be given to allow trapped partially sintered powder to be blasted away from small gaps, holes and mechanisms. Larger spaces may have to be included around complex parts to ensure the cake can be removed. In general, 1 mm clearance is usually sufficient to thermally isolate each part on the build platform.

10.2.6 Feature Type: Screw and Threads

Threads should always be built vertically, if at all possible.	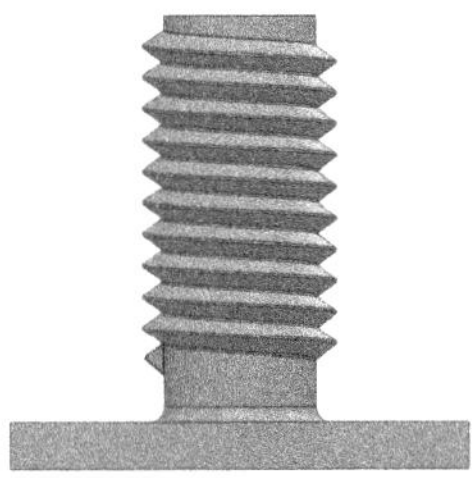

Comments
Because of the relatively rough surface finish of EBM, all threads will need to be tapped/machined.

Fillet the bottom of the boss where it meets the wall to avoid stress concentrations. A good rule of thumb is to make the fillet ¼ of the thickness of the walls.

10.3 Designing for Metal Binder Jetting

Metal binder jetting is an AM process in which layers of metal powder are bound together with a binder. The process to make finished metal parts involves several steps.

Printing: The printer deposits thin layers of metal powder onto a print bed and applies a binder, or liquid glue, to each layer to create your 3D model.
Green state: Once the printing phase is complete, the model is in a fragile "green state". It's important to consider the design of the model to ensure it can be handled during this temporary but critical phase.
Detail sizes: The minimum feature size for metal binder jetting is around 2 mm. Though the process can print finer details than this, occasionally as small as 1 mm, the fragility of the parts makes them hard to handle without damaging smaller details. This applies to both features and text.
Debinding: The model is cured in a furnace to burn away the binder.
Blasting: The loose steel powder around your model is removed by blasting it with air. Care must be taken for the air jet not to damage delicate parts.

And then either of 2 options to create solid metal parts:

Sintering: The part is placed in a furnace and heated until the particles of metal sinter together.
Infiltration: Sprues and stilts, which are passages that allow the bronze to infiltrate, are added to the model. Since the model is still in its green state, and fragile, it is surrounded with sand, which acts as a temporary support material. The model is then infused with bronze at an extremely high temperature, producing your metal/bronze matrix part. The ratio of metal to bronze is usually around 60% metal/40% bronz (Fig. 10.1).

10.3.1 Shrinkage

Shrinkage is highly dependent on the geometry of the part.

As a rule of thumb, for parts with length up to 25–75 mm shrink between 0.8 and 2%, while larger parts have an estimated average shrinkage of 3%.

Non-uniform shrinkage can be an issue and must be accounted for during the design stage in collaboration with the Binder Jetting machine operator.

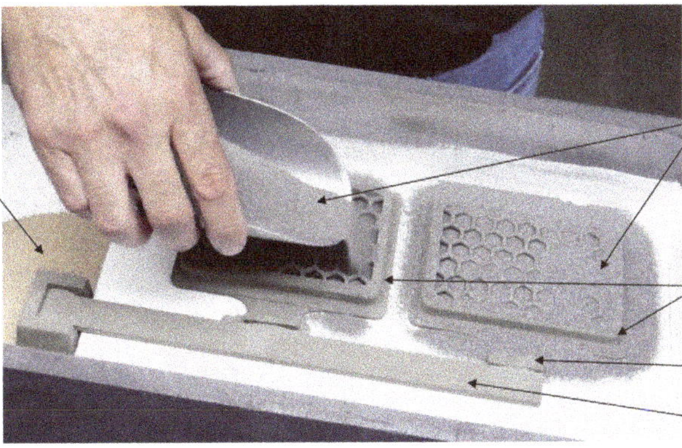

Fig. 10.1 The metal binder jetting bronze infiltration process. Courtesy of ExOne

10.3.2 Part Density

Binder Jetting metal parts will typically have some internal porosity. Bronze infiltrated parts will typically be 90% dense. Sintered stainless steel parts will typically be 97% dense.

This means that their mechanical properties will be inferior to the bulk material or to parts that were manufactured with a different 3D printing process, such as powder bed fusion. Material properties of the produced parts are roughly equivalent to metal parts produced with Metal Injection Molding

A comparison of the material characteristics of metal binder jetting parts is shown in the table below.

	Binder Jetting Stainless Steel 316 (sintered)	Binder Jetting Stainless Steel 316 (bronze infiltrated)	DMLS/SLM Stainless Steel 316L
Yield strength (MPa)	214	283	470
Elongation at break (%)	34	14.5	40
Modulus of elasticity (GPa)	165	135	180
Surface roughness without post-processing (μm)		Ra 15	Ra 15–50

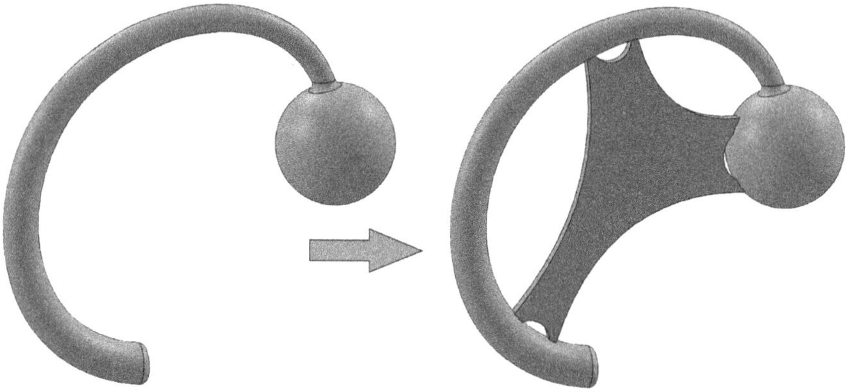

Fig. 10.2 Making the part strong enough to handle in its green state

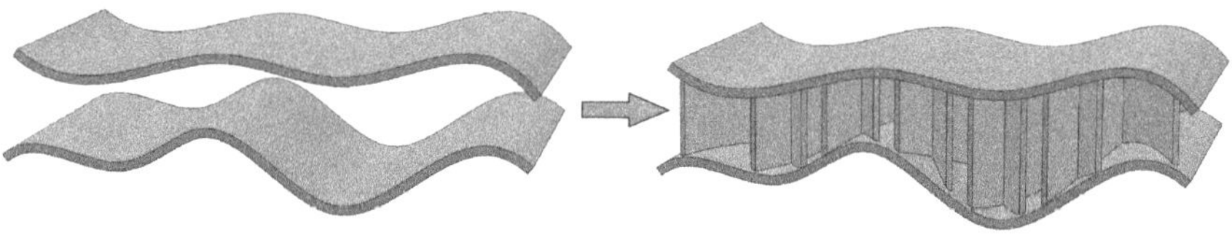

Fig. 10.3 Avoid large surface area unsupported walls

10.3.3 The Most Important Design Rule for Metal Binder Jetting

Green parts are fragile. Everything in your design is influenced by this.

Until the part has been infiltrated with bronze, or sintered, it is fragile and poor design may result in the part being damaged during handling. Because of the temporary fragility of your model during the green state, certain geometries can't be supported. Your model may print without a problem, but you may not be able to handle it before it's hardened. It is therefore often necessary to include strengthening ribs or walls in your design that allow the part to be handled in its green state (Fig. 10.2).

This also applies to any large surface that is unsupported may need a geometrically stronger structure, like a lattice, to help it hold its own weight (Fig. 10.3).

Figure 10.4 shows and example of 3 different designs for the same part.

Sharp external corners and "knife edges" can chip and crack during depowdering, handling, or heating while your model is in the green state. Rounding off the edges with at least a 1 mm fillet will prevent your model from breaking. If you need to sharpen the edges, you'll be able to do so during your own post-production processes (Fig. 10.5).

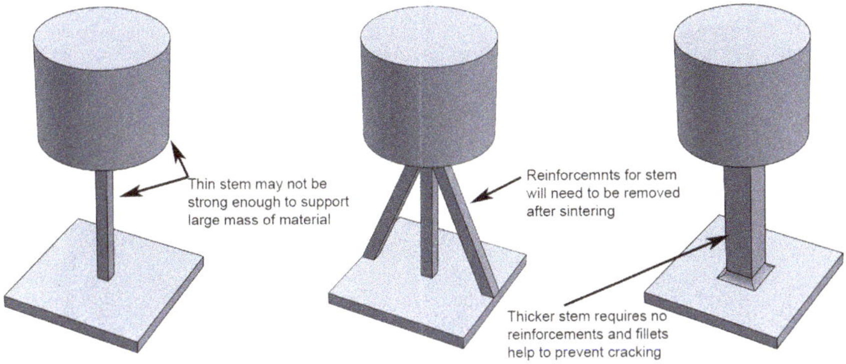

Fig. 10.4 Three different designs. The one on the right is the easiest for successful metal binder jetting

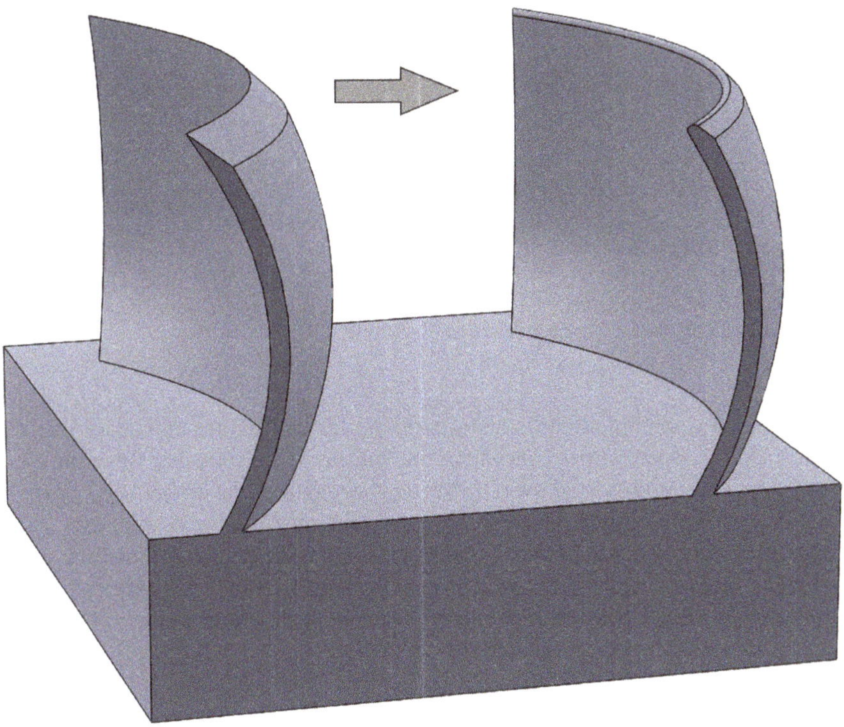

Fig. 10.5 Fillet all knife edges to avoid them breaking. In this design, the weak top sharp knife edge has been reduced by the fillet

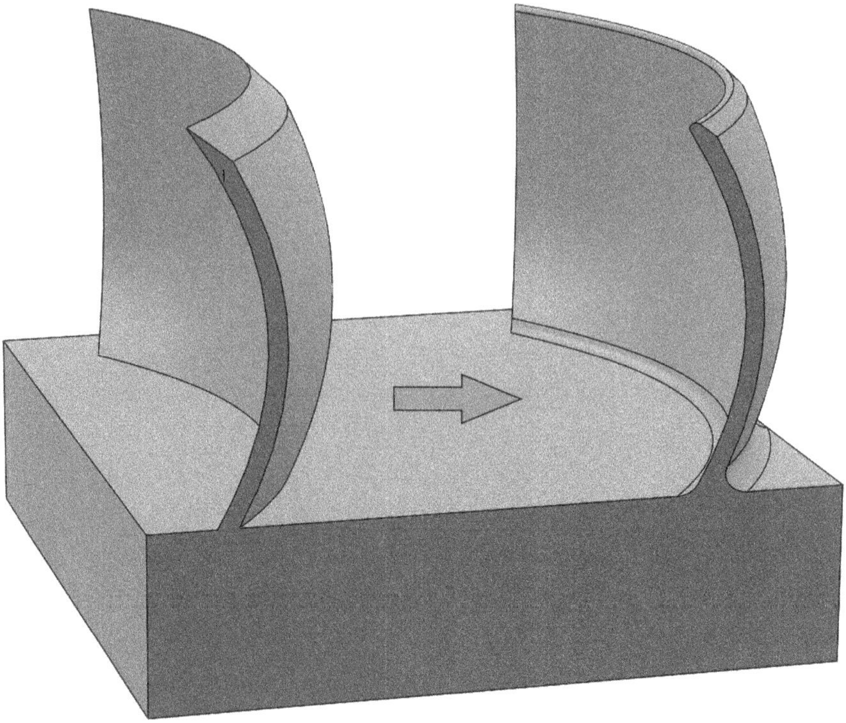

Fig. 10.6 Fillet all internal corners to avoid them becoming stress concentrations. In this design, the weakness at the junction has been removed by the fillet

Filleting or rounding edges is also beneficial for concavities and sharp internal corners. This is where stress concentrations will occur, so rounding them prevents these issues and improves the overall structural strength of the model (Fig. 10.6).

If the connection points between parts of your model are too thin, they can't properly be infiltrated. Models with abrupt changes in the amount of material (such as designs which go from very thick to very thin material volume) may also break during this process. Connection points in your model should be smooth transitions and provide sufficient surface material (Fig. 10.7).

Fig. 10.7 Ensure connection between parts is thick enough to allow infiltration between parts, and strong enough to handle

10.3.4 Feature Type: Wall Thicknesses

Build size	Minimum wall thickness	
3–75 mm (0.125–3 in.)	1.0 mm (0.04 in.)	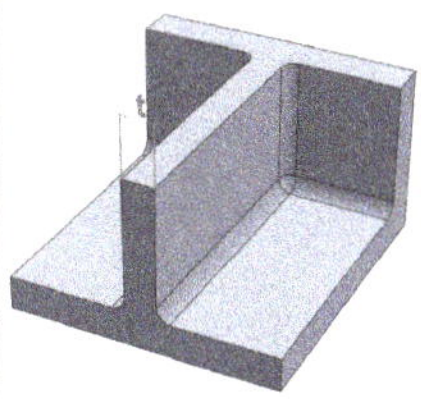
75–152 mm (3–6 in.)	1.5 mm (0.06 in.)	
152–203 mm (6–8 in.)	2.0 mm (0.08 in.)	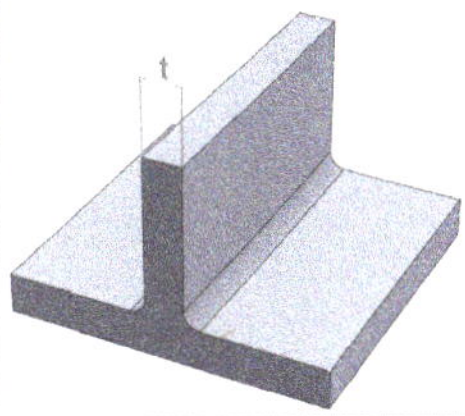
203–305 mm (8–12 in.)	3.2 mm (0.12 in.)	

Comments

The main criteria that determines the wall thickness is the ability to handle the part while it is in its green state. So shorter, well protected walls can be thinner than unprotected walls or walls of large surface area. Less supported walls must be thicker than well supported walls.

Fillet all internal corners with a radius of at least 1 mm.

10.3.5 Feature Type: Overhang

Minimum thickness (t)		
Thickness	>2 mm (0.08 in.)	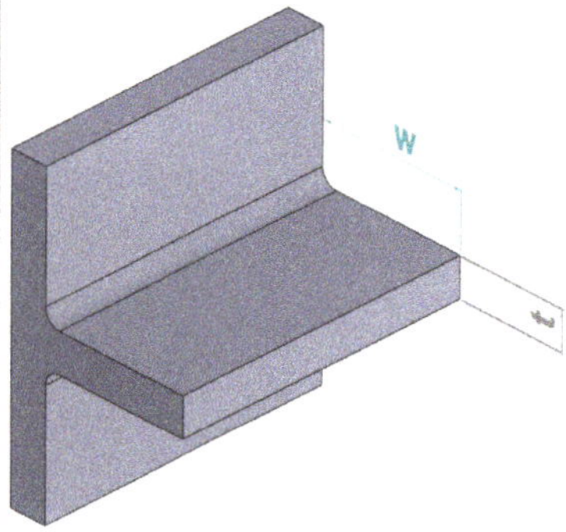
Width	25 mm (1 in.)	

Comments

The main criteria that determines the overhang dimensions is the ability to handle the part while it is in its green state. So shorter, well protected overhangs can be thinner than unprotected overhangs or overhangs of large surface area.

Fillet all internal corners with a radius of at least 1 mm.

10.3.6 Feature Type: Holes

Minimum diameter for circular horizontal hole (h)	Minimum diameter for circular vertical hole (v)	
2 mm (0.078 in.)	(0.06 in.) 1.5 mm	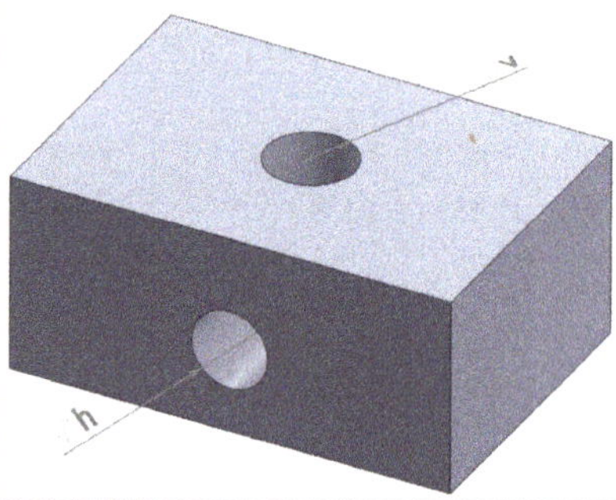

Comments

A minimum diameter of 1.5 mm vertically or 2 mm horizontally is recommended to ensure that rough surfaces do not cause the holes to close up, or that the infiltration material does not clog the holes. For holes through walls that are around 2 mm thick, holes as small as 1 mm can sometimes successfully be printed, depending on the part geometry.

In walls thicker than about 3 mm, vertical holes will generally need to be no smaller than 2 mm.

10.3.7 Feature Type: Salt-Shaker Holes

Minimum diameter for salt-shaker hole (d)
5 mm (0.2 in.)

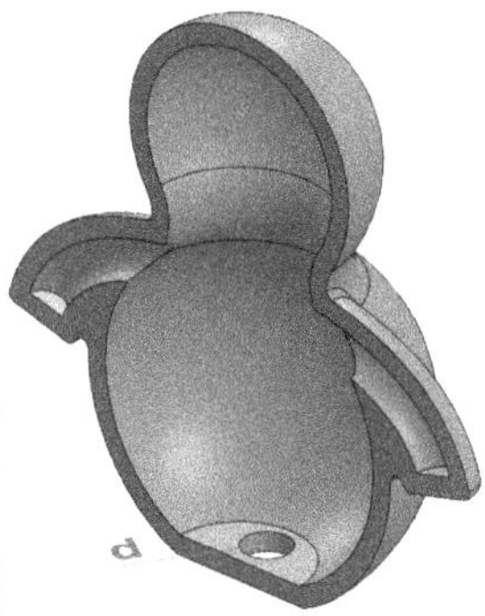

Comments

If you re producing entirely hollow components, they will be filled with lose powder. This loose powder will cause problems during infiltration and sintering, so it is best to remove it. To remove the loose powder after printing, you must include salt-shaker (or 'escape') holes in your design. Salt-shaker holes should be as large as possible to make the internal powder as easy to remove as possible. In general, a hole size of at least 5 mm diameter is recommended and at least 2 escape holes should be used so air can be blown into one hole and the powder can escape from the other. On relatively thin-walled parts, salt-shaker holes as small as 1 mm can occasionally be used, but powder may take longer to remove from inside the part.

Chapter 11
Other AM Considerations

This chapter covers activities that are, typically, outside the designer's influence, but impact the quality of the final part produced.

11.1 Designer Machine Operator Cooperation

One of the very important aspects that determines the quality of print jobs is the build orientation, as discussed in an earlier chapter. Certain design features are easier to build, have better strength characteristics, and/or require fewer supports if built at a particular orientation. Sometimes, conflicts can develop among the needs of different design features.

Consequently, close cooperation with the AM system operator is required to arrive at a satisfactory compromise. AM machine operators often have great amounts of tacit knowledge about what will, or won't, print well on the particular machine they are using, and about different strategies the designer can use to improve their part designs.

Another consideration is the build chamber size. If a part is too large, it must be sub-divided into smaller parts and assembled after the build. This may be a task taken care of by the designer or it might be assigned to the AM system operator.

Either way, the part should not be sub-divided through any critical features, and adequate location features must be added to ensure that it fits together correctly. Although usually employed through necessity, this approach can also be used to speed up or reduce the cost of building a large part, because the subdivided parts can sometimes be nested together to reduce the overall build volume, but might require excessive supports or build time.

Finally, the application of post-processing techniques will not usually be the designer's task, but could have a profound impact on the final outcome. Over-finishing of parts can further reduce the already variable dimensional accuracy of AM processes. The designer should communicate with whoever is responsible for

O. Diegel et al., *A Practical Guide to Design for Additive Manufacturing*, Springer Series in Advanced Manufacturing,
https://doi.org/10.1007/978-981-13-8281-9_11

post-processing to make sure that critical dimensions and/or features are preserved. If this proves to be impractical, the designer should specify additional operations to regain the required accuracy, e.g., machining of key features. Additional machining stock must be added to the CAD model to support this method.

11.2 Health and Safety

As the uptake of AM increases, the industry has started to realize that there are many health and safety aspects of AM that must be considered by companies running AM systems.

This applies to both polymers and metals.

With metals the safety of the machine operator is critical, and the main safety aspects to consider include: material exposure, gas monitoring, gas exhaust, material handling, and the risk of explosion.

11.2.1 Material Exposure

The use of metals in additive manufacturing is growing and so is the importance of understanding the risks of exposure to powdered metals. In the additive industry, the average particle size is between 25 and 150 μm, requiring special handling and storage. Metal toxicity is a real threat. The human body cannot readily metabolize most of these metallic powders, and build-up through exposure can quickly reach toxic levels. Protective gear and respiratory masks, of an adequate quality, should always be worn when handling metal powders, and any other potentially toxic AM materials.

11.2.2 Gas Monitoring

Any machine that uses laser melting runs with odourless and colourless gases that displace oxygen, usually Argon or Nitrogen.

Some companies install these machines on manufacturing floors that have plenty of open space but others are, instead, locating them in smaller rooms which can be problematic. Installing an oxygen sensor is always recommended wherever these machines are placed to continually record the level of oxygen in the room and issue a warning should oxygen levels drop below the safe minimum.

11.2.3 Gas Exhaust

Users need to consider how to handle exhaust gases. Some machines expel dangerous gases when they print, requiring ventilation to the outside of the building or into special gas emissions storage tanks.

Safety training should cover off-gassing and proper management of volatile organic compounds (VOCs) and inorganic compounds. Sometimes, the use of a charcoal scrubber is warranted too.

Companies should develop an air quality management plan with system vendors or consult with a professional organization to develop a risk-reduction strategy.

Even desktop polymer 3D printers can emit hazardous or malodorous gases, and should be used in well-ventilated areas or use some kind of fume extraction system.

11.2.4 Material Handling

Most manufacturing plants deal with bulk material. Metal AM deals with powders, and metal in powder form can be explosive. How do you store it? There is also some waste in the AM process and, although there is not a lot of powder waste, how do you dispose of it? Every company needs to develop delivery, handling, and storage procedures to complete their plant safety plan. Reactive metals, such as aluminium and titanium, for example, should not be stored in rooms where there is a water-based sprinkler system.

11.2.5 Risk of Explosion

Metal powder can be explosive! In processes involving powdered metal, static electricity is a concern. When a static arc can ignite the powder, the metal is considered reactive, and special care must be taken.

Because static electricity can lead to fire, the type of fire extinguisher that is readily available in the vicinity of an AM operation matters. Most companies already have Class A, B, and C fire extinguishers on hand, but AM metals, which are combustible, require a Class D extinguisher.

11.3 AM Part Certification

The introduction of additive manufacturing into critical applications present some challenges. This is because, for such critical applications, we need to be able to guarantee the quality and consistency of the parts produced.

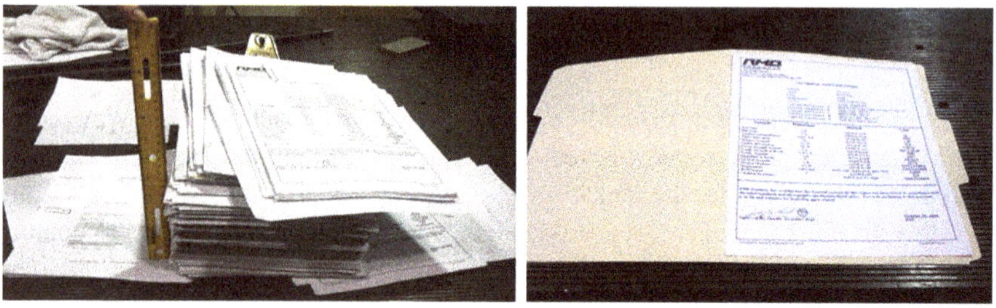

Fig. 11.1 Paper work for conventional manufacturing certification compared to paperwork for additive manufacturing certification. Courtesy of Chris Glock, RMB Products Manufacturing

Examples of critical applications can be found in commercial and military aviation, manned and unmanned space applications, and medical applications. Such applications often require some form of certification to establish suitability for service, and typically have stringent quality and traceability requirements.

With design for AM, part-consolidation allows many components to be consolidated into a single more complex component. From a certification point of view, this can mean certifying a single component compared to many components, and the processes required to join them (Fig. 11.1).

Preparing a model for printing may see several iterations of an object printed at various orientations, and print parameters, before the part comes out of the machine within design specifications. In the case that a design needs to be changed over to a different machine type, this may have to be performed all over again, given changes in size, energy type and more.

11.3.1 What Needs to Be Certified?

Depending on the type of parts, and the standards being applied, there are a number of elements that can form part of the AM part certification process including:

- The company.
- The material.
- The machine.
- The print location on the build platform.
- The print orientation.
- All the print parameters.
- The post-processing.
- Sometimes test coupons must be printed for destructive verification.
- Upon qualification, each of the production parameters is fixed, with any deviations requiring additional testing.

There are, generally, three different paths to qualification:

- Statistically-based qualification, based on extensive and expensive empirical testing.
- Equivalence-based qualification, achieved through moderate testing to demonstrate a new material or process is equivalent to a previously qualified material or process.
- Model-based qualification, where a material's or process' performance is demonstrated in a computer model and demonstrated with minimal testing. For AM, this technology is still not mature.

There are currently some existing certification guidelines on the market that help in providing confidence to manufactures and end-users that parts are safe, reliable and robust. Certification schemes such as the Lloyd's Register's Additive Manufacturing Product Scheme have clients demonstrate that their parts meet existing manufacturing methods and standards using their Guidance Notes for Additive Manufacturing of Metallic Parts. Companies that pass these tests are awarded an LR certification, meaning end-users know that whatever parts they order are safe for their particular application.

Chapter 12
Post-processing

All Additive manufacturing (AM) technologies require post-processing to produce parts that are ready for use. This post-processing can range from support material removal, to surface quality improvement, to colouring and painting, and to aging for polymer parts and heat-treatment for metal parts. Throughout the AM industry there is a vast amount of tacit knowledge in the area of post-processing but there, currently, exists very little documentation on the various post-processing methods for different AM technologies and materials. This leads to time being wasted by companies having to individually learn and develop post-processing methods. This chapter aims to correct this.

The overall process flow of additive manufacturing includes pre-processing and post-processing, and is presented in the following table. Note that the steps can vary, sometimes greatly, depending on the application, material, AM system being used, and specific requirements of the parts.

Metal powder-bed fusion	Polymer powder-bed fusion	Material extrusion	Vat photopolymerization	Binder jetting
Check quality of files and repair if necessary	Check quality of files and repair if necessary	Check quality of files and repair if necessary	Check quality of files and repair if necessary	Check quality of files and repair if necessary
Prepare print-job in software by arranging parts on build platform and generate support material	Prepare print-job in software by arranging parts on build platform	Prepare print-job in software by arranging parts on build platform and generate support material	Prepare print-job in software by arranging parts on build platform and generate support material	Prepare print-job in software by arranging parts on build platform
Clean AM system	Clean AM system	Clean AM system	Clean AM system	Clean AM system

(continued)

O. Diegel et al., *A Practical Guide to Design for Additive Manufacturing*, Springer Series in Advanced Manufacturing,
https://doi.org/10.1007/978-981-13-8281-9_12

(continued)

Metal powder-bed fusion	Polymer powder-bed fusion	Material extrusion	Vat photopoly-merization	Binder jetting
Preheat build chamber	Preheat build chamber	Preheat build chamber		Preheat build chamber
Print	Print	Print	Print	Print
Remove build plate from build chamber	Find and remove parts from powder bed	Remove parts from build chamber	Drain and/or recycle unused material as applicable	Find and remove parts from powder bed
Remove loose powder and recycle	Recycle remaining powder	Remove support material	Remove parts from build chamber	Recycle remaining powder
Thermal stress relief	Media-blast parts to remove surface powder	Surface finish: sand, vapor smooth, paint, etc.	Remove support material	Air-blast parts to remove surface powder
Remove parts from build plate	Surface finish: tumble, sand, dye, paint, etc.	Inspect	Post-cure in UV chamber	Bake parts as necessary
Hot isostatic pressing	Inspect		Surface finish: sand, vapor smooth, paint, etc.	Strengthen with infiltration
Remove support structures			Inspect	Surface finish, sand, paint, etc.
Heat treat				Inspect
Surface machining, shot-peening, abrasive flow machining, etc.				
Inspect				

12.1 Support Material Removal

12.1.1 Polymer

Material Extrusion

There are 3 main forms of support material used in material extrusion technologies:

- The same polymer as the printed part is used as support: With these systems, a lower density version of the same polymer as the part polymer is used as support

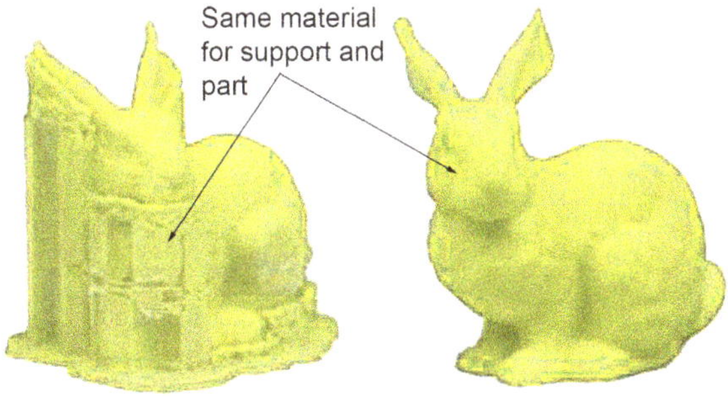

Fig. 12.1 Part printed with the same support material as the part

Fig. 12.2 Part printed with break-away support material

material. The support also makes only point contact with the part. The support material for these system is removed through mechanical force using hand tools. This form of support material is most commonly found on desktop 3D printers (Fig. 12.1).

- Break-away support material: This support material is a different polymer to that of the part. The support material for these system is removed through mechanical force using hand tools (Fig. 12.2).
- Soluble support: This support can be dissolved using an appropriate solvent. It can, however, take several hours to dissolve, particularly if there is support material in long narrow pipes into which the solvent can only dissolve small amounts at a time. This support is the best to use if the part has fragile features that would, otherwise be damaged if removed mechanically (Fig. 12.3).

Vat Photopolymerisation

Vat photopolymerisation systems use the same polymer for the support as is used for the printed part: With these systems, the support material generally takes the form of a 'tree' in which the tree branches are what make contact with the part and support it. The support also makes only point contact with the part. The support material for these systems is removed through mechanical force using hand tools.

Fig. 12.3 Part printed with soluble support material, courtesy of Joris Peels

Most vat photopolymerisation systems will require the part to be fully cured in a UV oven, or ambient sunlight, after it is removed from the printer. It is, generally, best to remove the support, and sand the part, while it is in between the two stages, and the material is still slightly soft so easier to remove (Fig. 12.4).

Material Jetting

Material jetting systems use a wax-like material as support material. Depending on the system, this support material can be washed off with a water jetting system, or melted off with heat (Fig. 12.5).

Powder Bed Fusion

Polymer powder bed fusion technologies are one of the few processes that do not require support material as the unmelted powder that surrounds the melted powder acts as support. When the printing process is finished and cooled, the parts are dug out of the powder 'cake' and blasted with air and sand (or powder) to clean off the part. The process is akin to an archaeologist digging parts out of a dig site (although the powder bed fusion powder is substantially easier to remove that it would be to remove clay or dirt from an archaeological dig site).

When sand-blasting, care should be taken not to get too close to the sand blasting nozzle, otherwise the sand can slightly 'burn' the part and leave brown marks behind. Laser sintered parts are particularly susceptible to this because they are usually made from white powder. If coloured powder or fusing agent is being used, such as in the multi jet fusion process which produces dark grey parts, this is much less of an issue.

By blasting the part with the polyamide powder itself, instead of sand, this problem can be avoided, but the parts may take fractionally longer to clean because the polyamide powder is less abrasive than sand (Fig. 12.6).

Because unsintered polymer powder gets 'caked' together during the printing process, powder can be hard to remove from long thin tubes or holes. A useful trick for making this easier is to leave cleanout holes along the length of the tubes for powder removal. The powder can then easily be blown out of the pipes with an air gun (Fig. 12.7).

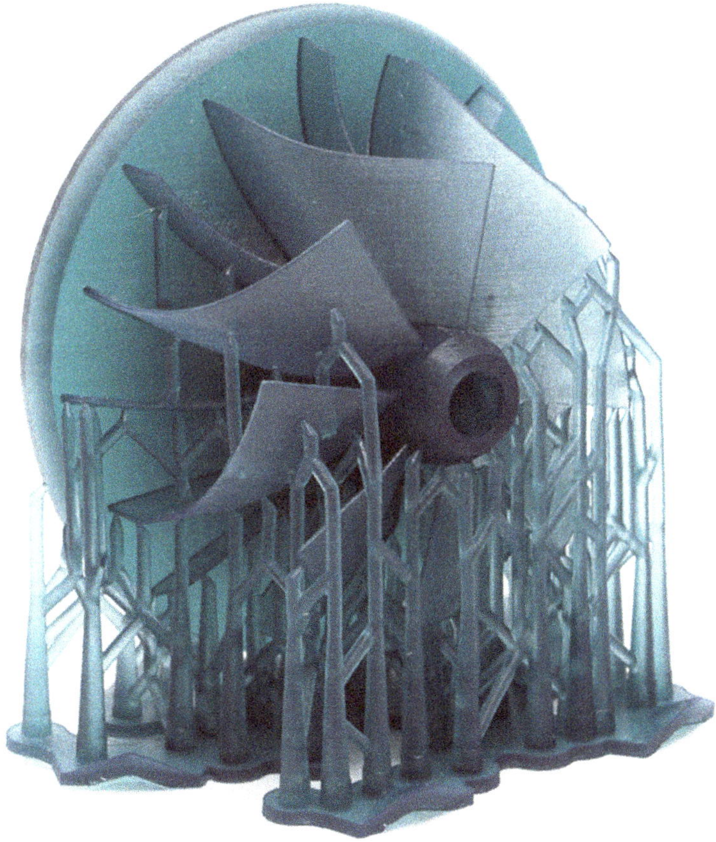

Fig. 12.4 Support material on vat photopolymerisation part, courtesy of www.3dhubs.com

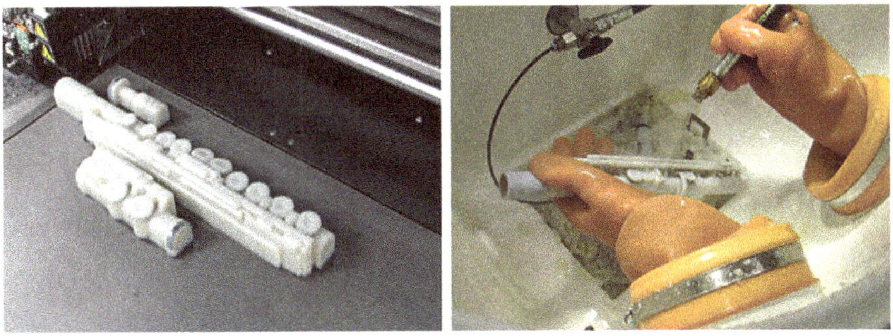

Fig. 12.5 Material jetting parts with water-blastable wax support material, courtesy of Zoran [6]

Fig. 12.6 Powder bed fusion parts being removed from the powder 'cake'

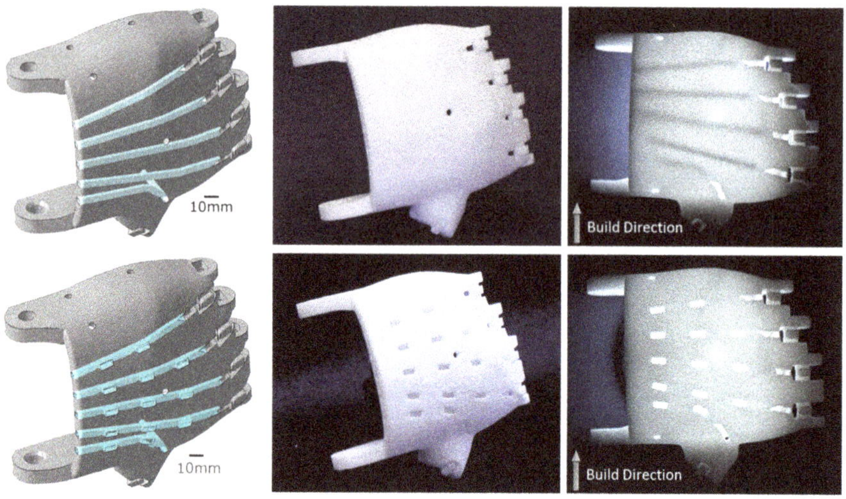

Fig. 12.7 Part designed with cleanout holes to make depowdering of the part easier, courtesy of Ben Weiss

12.1.2 Metal

Support material removal for metal parts can be challenging and require substantial amounts of time. This is why it is so important to design metal parts for AM to reduce the amount of support material required. To repeat a point made earlier, printing in metal is not easy! It requires considerable work before and after the part is printed, as well as a good understanding of the printing process itself.

The bracelets below, for example, make a nice bit of marketing for what metal AM is capable of. But it is not until one understands the work behind them, however, that one understands just how hard it can sometimes be to justify the economics of AM (Fig. 12.8).

The time to print one of the above bracelets is around 10 h per bracelet. The reason for this relatively long print time is that the bracelets are printed at an angle of 45° to

Fig. 12.8 Aluminium bracelets printed as single assemblies with moving parts

improve the bottom face surface finish and reduce the risk of stress cracks but this, unfortunately, increases the total build height and, therefore, the print time. If they are printed horizontally, the print time per bracelet would drop to approximately 3 h (Fig. 12.9).

The table below shows the time to take the bracelets through the various post-processing steps it must go through before being of an acceptable quality (Fig. 12.10).

After the print has finished
With metal AM the entire part comes out of the machine welded to the build plate. The first step is to heat treat the part to remove residual stresses, otherwise the part will distort when removed from the platform. This is discussed in greater detail in the section on heat treatment.

Removing the part from the platform
After heat treatment, the first step is to cut the part off the build plate. This is generally done with wire cutting (wire Electrical Discharge Machining, or EDM) or with a saw (Fig. 12.11).

It is for this reason that, when setting up a print job, the part is usually positioned either 2 or 5 mm above the build plate. If one is removing the part with wire EDM, 2 mm is required to account for the EDM wire thickness and, if one is using a saw, then 5 mm is needed to account for the thickness of the saw blade (Fig. 12.12).

In general, it is faster to remove the part with a saw, but it is more precise to use wire EDM. Wire EDM also has the advantage that it can be used to improve the quality of the bottom surface of the part while it is cutting. Or it can be used to surface the top face of the build plate so it can be re-used with minimal effort, whereas using a saw will require the top face of the build plate to be machined flat.

Another consideration, however, in the choice between using a saw or wire EDM is whether there is any loose powder trapped inside the part. As wire EDM uses a

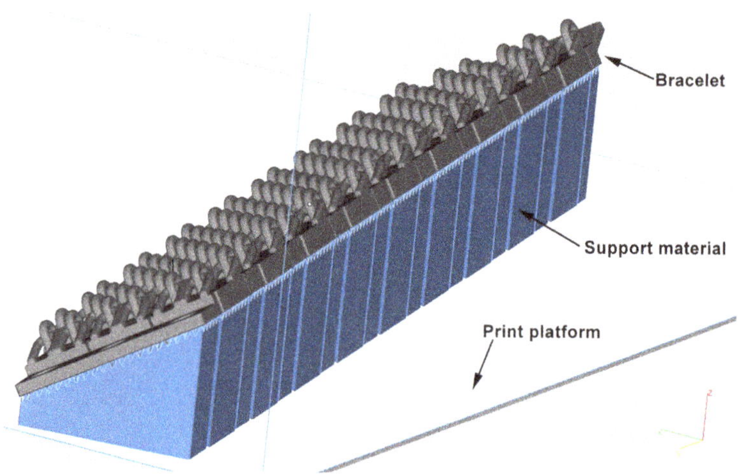

Fig. 12.9 Bracelet print orientation for best surface finish

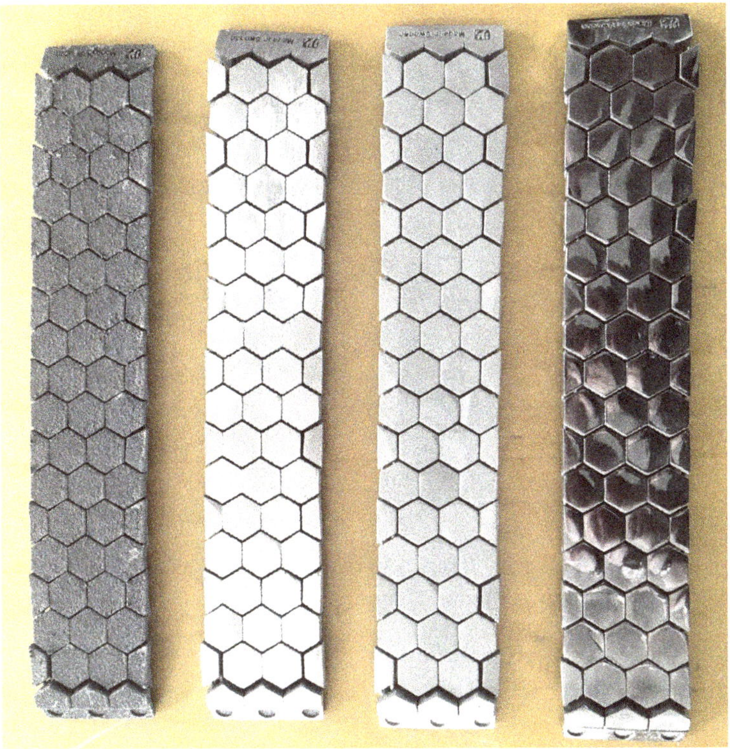

Remove from build platform and remove supports	Linish surface with belt sander	Media-blast with glass beads	Polish
30 minutes	30 minutes	5 minutes	4~5 hours

Fig. 12.10 Progression of different post-processing steps for bracelet

dielectric liquid as part of the cutting process, if a part has any internal channels, for example, that are still filled with powder, then the liquid is likely to infiltrate the powder inside the part, turning it into a paste, and making it very difficult to remove. So, if wire EDM is being used, it is best to ensure that all loose powder from inside the part has been removed before beginning the EDM process.

Removing the support material

Then begins the often arduous task of removing support material. If a part is well designed, and only has support material on the bottom, then it can quickly be removed through wire EDM, cutting, or grinding. In most cases however, the rest of the support

Fig. 12.11 Aluminium guitar still welded to build plate after coming out of the printer, having the loose powder vacuumed away, and being heat treated

material will need to be removed by hand, and this can be a time consuming process. The support material for metal AM parts is removed through mechanical force using hand tools. Considerable force is sometimes required to break away the support material (Fig. 12.13).

After all the support material has been removed, the part must then be given an appropriate surface finish, through sanding, grinding, shot-peening, machining, etc. (Fig. 12.14). This is discussed further in the next section.

The following table shows the time required for each pre- and post-processing task required to produce the above aluminium guitar.

Task	Time (h)
File preparation	2.5
Machine preparation	2
Printing	9
Machine cleaning	2
Stress relief	3
Cooling	30
Removal from build plate	15
Support removal	4
Surface treatment (filing, sanding, shot-peening)	4

Fig. 12.12 Aluminium guitar being cut off the build plate with a saw

12.2 Polymer Surface Treatments

12.2.1 Vapour Smoothing

Vapour treatment is a way of smoothing the surface of a print with a solvent that can dissolve the material of the print. This is done by vaporizing the solvent, and letting the vapour dissolve the outer surface of the part enough to make the layer lines disappear. For ABS acetone is used as a solvent and chloroform can be used for PLA. **Great care should be taken when working with these chemicals** (Fig. 12.15).

By heating the fluid, a vapour with little drops of the solvent will be formed. This will slowly settle on the print and start to dissolve the outer layer of plastic. When timed correctly, you are able to get a very smoothly finished print, similar to an injection molded part. It should be noted, however, that, as you are dissolving the outer surface of the part, dimensional accuracies will be affected by this process.

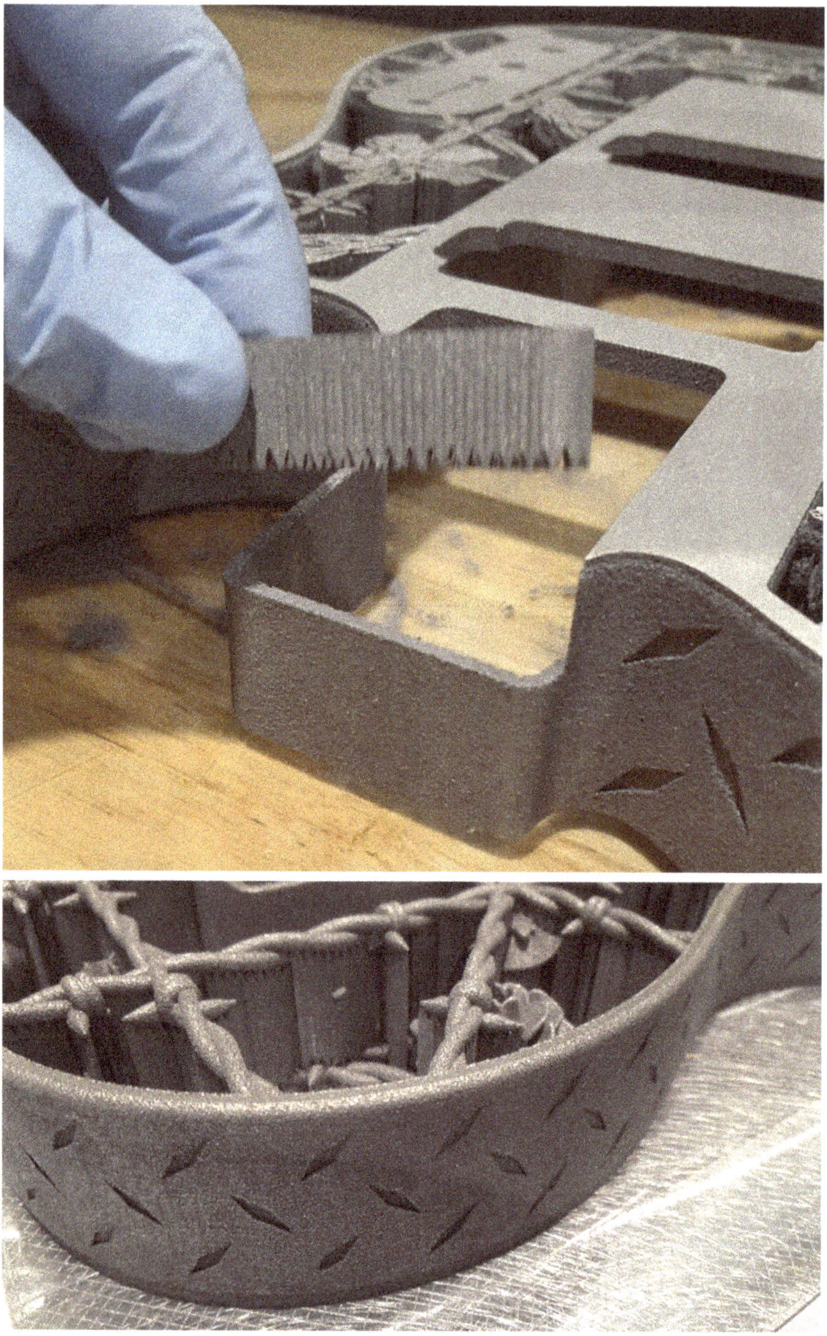

Fig. 12.13 Support material that must be removed from aluminium guitar

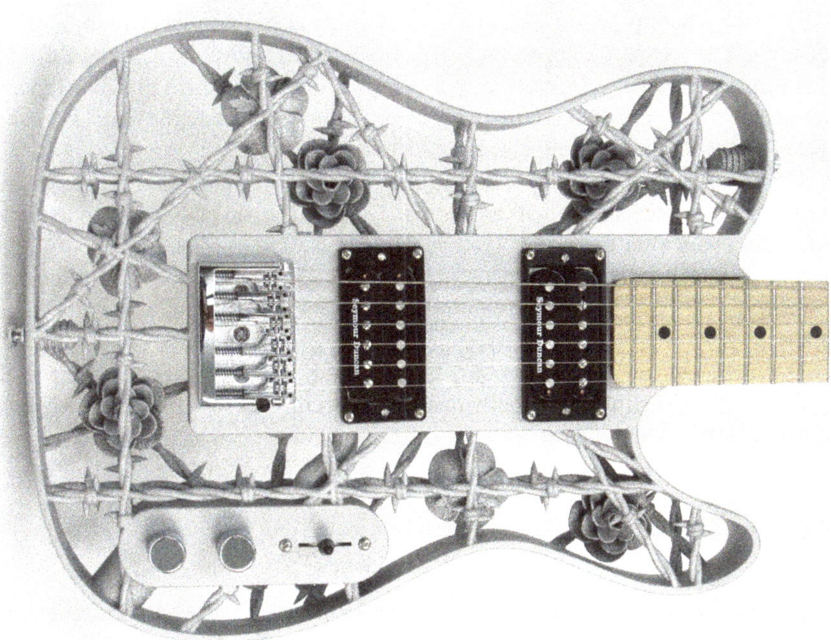

Fig. 12.14 AM manufactured aluminium guitar

Fig. 12.15 Acetone vapour smoothed ABS parts, courtesy of Andrew Sink

The simplest way to achieve vapour smoothing is to use a jar, with a bit of the solvent in the bottom of the jar, and suspend your part above the solvent. Even just doing this in ambient temperature will work, as the solvent will slowly evaporate and coat the part and dissolve its outer layer. Using a hot plate, or the 3D printer's heated bed, will greatly accelerate the process.

Timing of vapour treatment is critical, as the longer the part is left in the vapour, the more the plastic surface will be melted. The melting can also keep occurring for several minutes after the part is removed from the vapour. Exact timing is hard to specify, as it very much depends on the brand of ABS or PLA being used and the geometry of the part. The correct timing is best established through some timed trials of test parts.

The advantage of using a vapour, as compared to just dipping the print in the solvent, is that the solvent is equally distributed over the print. This will give a more consistent result. Material extrusion processes are also somewhat porous, so dipping a part can soak the solvent into the part so it will keep melting the part long after it has been removed from the liquid solvent.

Using fluids such as acetone and chloroform is risky; make sure to only use this process if you know how to work with these materials!

- Acetone vapours can cause irritation or muscle weakness and are highly flammable, so keep it away from fire.
- The vapours released by chloroform can be irritating to the eyes/skin or respiratory tract. They may also cause dizziness.
- Always dispose the leftover solvents in a chemical box.

12.2.2 Tumbling

Tumbling, also known as rumbling, is a type of mass finishing manufacturing process used to deburr, radius, descale, burnish, clean, and brighten AM parts. This process is applicable to most AM technologies.

In this batch-type operation, specially shaped pellets of media and the AM parts are placed into the tub of a tumbler and rotated or vibrated. There are many types of tumbling systems including vibratory tumbling, rotary tumbling, barrel finishing, centrifugal barrel finishing, etc. The movement action causes the media to rub against the AM parts, which gradually wares down the part surface and yields the desired result. Depending on the application, this can be either a dry or wet process.

Tumbling is a time dependent process where the longer the parts are left in the tumbler, the more worn (smooth) the parts will be come. Typical tumbling times, for LS parts, range from 3 to 6 h depending on the abrasive medium used (Fig. 12.16).

It should be noted that tumbling is an abrasive process, so part accuracy can be affected as the parts are, effectively, being worn. Sharp corners will also become slightly rounded.

Fig. 12.16 Examples of tumbled AM parts, courtesy of Duann Scott

Fig. 12.17 Example of dyed part sample colour swatch, courtesy of Midwest Prototyping

12.2.3 Dying

Dying is a particularly good technique for applying colour to powder bed fusion polymer parts. Almost any synthetic material clothes dyes, or leather dyes, can be used. In most cases, it is just a matter of following the instructions for each particular type of dye (Fig. 12.17).

The hue of colour largely depends of how long the parts are left in the dye. For most synthetic clothes dyes, it takes about 45 min to achieve a good colour with the dye bath at between 80 and 100 °C (176–212 °F) stirring continuously. The longer the part stays in the dye, the darker it gets. It is also important to constantly stir the dye, otherwise the parts end up with patches of darker and lighter colour on them.

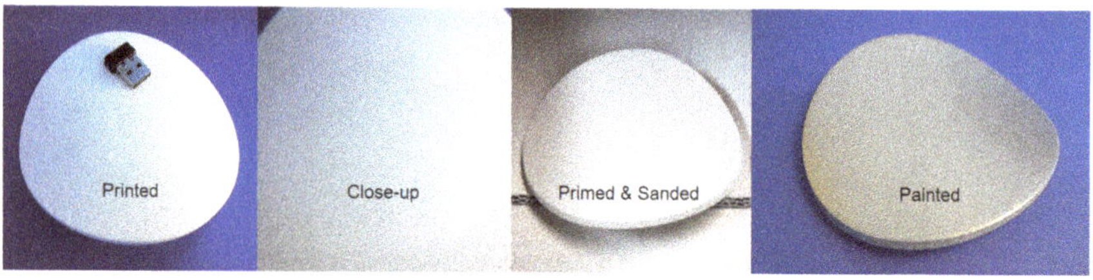

Fig. 12.18 Example of painted powder bed fusion process parts. But the same process applies to any other AM technology, courtesy of Antu Gortari and Sculpteo

12.2.4 Painting

Painting applies equally to all AM technologies and is one of the most common surface treatment processes for polymer AM parts. The process is very much the same as for any other form of painting.

The surface must first be prepared as much as possible by sanding. As with any other form of surface preparation for printing, one often starts with a relatively coarse grit san paper, around 120, and then moves up to gradually finer sand paper going through 240, 400 and then 800 grit papers. As many coats of sanding primer as is necessary to achieve a smooth surface are then applied, with sanding operation between each coat. The better the surface is prepared, the better the final results will be. In some cases, if the surface finish of the part is very rough, or has a very pronounced 'stair-step' effect, it can be quicker to apply a coat of automotive body filler to the part, and sand that down before applying a few coats of sanding primer.

Once the surface is smooth, the final colour is applied, again often requiring several coats, and is followed by a few coats of clear-coat both to protect the part, and to give it the appropriate desired level of glossiness.

Acrylic automotive paints generally produce excellent results. Some AM technologies, such as most material extrusion technologies will require more sanding and primer coats than others to achieve a smooth surface finish. Using matt, or stain, clear-coat as the final layer is a good technique to replicate the surface finish of many injection molded components. Gloss clear-coats can, of course, also be used, but may require extra cutting and polishing, with standard automotive compounds, to achieve a mirror-smooth finish (Fig. 12.18).

12.2.5 Using Textures

Applying 3-dimensional textures, such has leather, shark skin, woven pattern, etc. to the surface of a part can, surprisingly, hide much of the stair-stepping effect that is very clearly visible without the texture. Even on very gently curving parts,

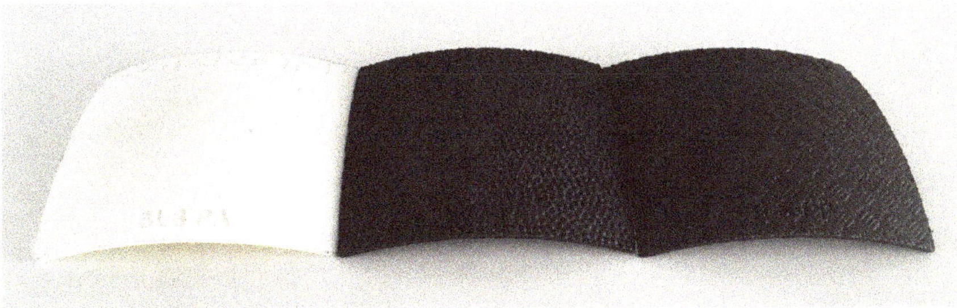

Fig. 12.19 Texturing the curved surface of a part almost entirely eliminates the stair-step effect. The left part clearly shows stair-stepping, whereas the textured parts do not

texture can make the stair-stepping disappear almost completely. This technique works particularly well on polymer powder bed fusion parts (Fig. 12.19).

There are now several software packages (Materialise 3-matic, Z-Brush, etc.) that allow true 3D textures to be added to the surface of a part, but care must be taken that the file size does not become too large to handle particularly if it needs to be converted into an STL file.

12.2.6 Sand Blasting

Abrasive sand or media blasting is the operation of forcibly propelling a stream of sand against a surface under high pressure to smooth a rough surface, roughen a smooth surface, shape a surface, or remove surface contaminants. In the context of additive manufacturing, it is mainly used for powder bed fusion as a technique for removing the powder that is stuck to the surface of the part.

As sand-blasting is an abrasive process, care should be taken not to get too close to the sand blasting nozzle, otherwise the sand can slightly 'burn' the part and leave brown marks behind. A general recommendation is to keep the part approximately 30 cm (12") from the sand blasting nozzle. In many cases, the sand can be replaced by used polyamide powder. By blasting with the polyamide powder itself, the burn problem that can occur with sand can be avoided, but the parts take fractionally longer to clean.

12.2.7 Machining

In many cases if an engineering quality surface finish or accuracy is required, then manual or CNC machining may be the only way to achieve this. The machining process is exactly the same as for any other polymer machining, except that greater care, or slower machining speeds, may be required for some of the AM processes

Fig. 12.20 Artwork of sound waves converted to 3D, by artist James Charlton, printed in gypsum with binder jetting, painted with conductive paint, and electroplated in antique silver

with higher degrees of anisotropy (weakness between the layers), such as those from material extrusion technologies.

12.2.8 Metalizing

All conventional polymer metalizing techniques that are used for conventional plastic parts can also be applied to AM parts. These include electroless plating or electroplating, and vacuum metalizing or PVD. If electroplating is used, or any other process that requires the part to be conductive, then a conductive paint must first be applied to the part.

Metalizing can produce parts that genuinely look and feel like metal parts, as they are genuinely coated with metal, except that they are noticeably lighter than their metal counterparts (Fig. 12.20).

12.2.9 Wrapping

Wrapping a part consists of covering it with a stretchable polymer film. This technique is commonly used in the automotive industry, but can be used with AM parts, assuming the part is not too complex. Wraps can be textured to add a 3D effect to the part.

The only preparation that needs to be taken before wrapping is to endure that the material surface is smooth enough for any stair-stepping not to show through the wrap material, and that the surface is clean and dry so that the wrap can adhere to the surface.

12.2.10 Hydrographics

Hydrographics, also known as hydro dipping, immersion printing, or water transfer printing, is a method of applying printed designs to three-dimensional surfaces. The pattern or image to be applied is first printed onto a soluble PVA film which is placed onto the surface of a water tank which dissolve the film and leaves the images ink floating on the surface of the water. The part is then carefully lowered onto the ink layer floating on the surface of the water and pushed into the water to transfer the image ink onto the part.

12.3 Metal Surface Treatments

The surface finish of metal AM parts is often too rough for some applications, so it is necessary to improve it. Surface roughness is determined by the AM process used, particle size of the material, layer thickness, build orientation, and the presence of supports.

It can be difficult to specify part roughness achievable with metal AM, particularly as the top, bottom, angled, and vertical surfaces each have a substantially different roughness. Laser-based powder bed fusion processes commonly produce as-built parts with a surface finish of about 7–15 μs (300–600 μin.) Ra on the top-facing and vertical surfaces. With Arcam's EBM process, the surface finish can be 20–25 μs (800–1000 μin.) Ra. Down-facing surfaces, and surfaces where support material is attached can, however, be substantially rougher, and on the bottom horizontal surfaces, where support material is usually attached Ra values as rough as 1000 μs can often be measured.

A variety of processes can help reduce the surface roughness of metal AM parts. Some involve mechanical action (e.g., machining, shot peening, and tumbling), while others involve chemicals combined with some type of mechanical action (e.g., electro-polishing). Each method must be evaluated based on how well it works, how much material is removed, cost, and the level of finish required.

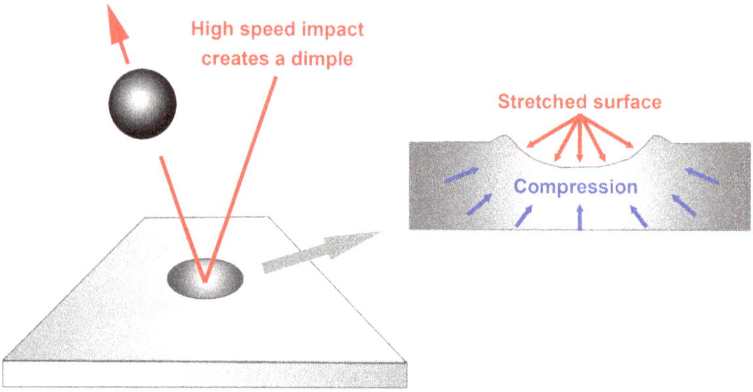

Fig. 12.21 Shot-peening has a micro-forging effect that can strengthen the surface of a part

12.3.1 Shot-Peening

Most metal AM parts are subjected to media-blasting (usually with sand or glass beads) as the first post-processing step after the support material has been removed from the part. This process helps to clean the part as well as remove any residual powder that is still attached to the part.

Shot-peening is a process, similar to sand-blasting, that uses compressed air to shoot small spherical particles against the part. Whereas sand-blasting is an abrasive processes that remove material from the surface, shot-peening flattens all the peaks of the rough surface that protrudes from the part, and has a micro-forging effect on the part which, not only smooths the surface, but can also strengthen the part (Fig. 12.21).

Common media for shot-peening includes glass beads and steel ball-bearings.

Most metal AM parts are subjected to shot-peening as the first post-processing process after the support material has been removed from the part.

12.3.2 Plasma Cleaning and Ion Beam Cleaning

Plasma cleaning is the process of removing matter from the surface of an object through the use of an ionized gas called plasma. This is generally performed in a vacuum chamber utilizing gases such as argon and oxygen, as well as mixtures such as air and hydrogen/nitrogen. The plasma is created by using high frequency voltages (typically kHz to MHz) to ionise the low pressure gas (typically around 1/1000 atmospheric pressure), although atmospheric pressure plasmas are now also common.

Ion-beam cleaning technology can be used for finish cleaning of the surface of AM parts with accelerated ion beams with energy of up to 1500 eV from molecular particles, adsorbed gases such as argon, polymer fragments, and water vapours.

12.3.3 Machining and Grinding

With metal AM, in general, the down-skin (the surface that is on the bottom surfaces of the part) and any areas that make contact with support material can have a rough surface finish. The top surfaces can also have patterns on it that are left behind by different laser hatching strategies. These surfaces must often be improved through filing, grinding or sanding.

If better surface finishes than those provided straight off the AM system are required, such as those required for a gasket to seal, or engineering accuracies, such as those required to press-fit a bearing, then the part must be machined as a secondary operation. Machining an AM part is no different from machining any other metal part. It gets clamped into the mill or lathe and is machined using the same speeds and feeds as for the conventional material.

Some ways to improve the machining of AM parts includes:

- Don't forget to add some extra material to those surfaces that will require machining. Usually around 0.5 mm extra material is enough.
- Add fixtures and mounting points to the part to make it much quicker to mount into the CNC machine. Often, what takes the longest in this secondary operation, is the time it takes to mount the part in the CNC machine. The cutting is often just a finishing cut in which around 0.5 mm of material is removed from the surfaces where high accuracy or good surface finishes are required. So any features that can make the part faster to mount into the CNC machine, and set the machine origin, will reduce the time needed for the overall operation.
- Counterintuitively, make the surfaces where you need the best surface finish those where you put the support material, as these surfaces are the ones that are likely to need to be machined anyway.

12.3.4 Abrasive Flow Machining

Abrasive flow machining, or extrude hone, is a technique for polishing internal channels by pumping an abrasive paste through the internal channels in the part. The abrasive particles in the media grind away rather than shear off the unwanted material.

The rate of material removal depends on the following factors:

- Media flow rate
- Viscosity
- Abrasive particle size
- Abrasive concentration
- Particle density
- Particle hardness
- Workpiece hardness.

12.3.5 Anodizing

Anodizing is used to produce protective and decorative oxide layers on aluminium, improving corrosion protection and wear resistance. Different colours can be created by dying or electrolytic colouring.

Aluminium AM parts can be anodized in exactly the same way as conventional aluminium parts. With AM parts, however, you may have more freedom to design in better hanging points for when the part is hung in the anodizing bath.

12.3.6 Plasma Spraying

Plasma spray is a thermal spray coating process used to produce a high quality coating by a combination of high temperature, high energy heat source, a relatively inert spraying medium, usually argon, and high particle velocities. Plasma is the term used to describe gas which has been raised to such a high temperature that it ionizes and becomes electrically conductive.

The utilisation of plasma spray coating technology allows the spraying of almost any metallic or ceramic on to a large range of materials with exceptional bond strength, while minimising distortion of the substrate.

Plasma spraying can be used to improve:

- corrosion protection
- wear resistance
- heat and oxidation resistance
- temperature management
- electrical resistivity and conductivity.

12.3.7 Plating and PVD

Electroplating can be applied to AM parts in the same way as it can be applied to conventionally manufactured parts. Electroplating is also known as electrodeposition because the process involves depositing a thin layer of metal onto the surface of a work piece, which is referred to as the substrate. An electric current is used to cause the desired reaction.

How electroplating works if a layer of, say, nickel is to be electroplated onto a metal AM part to improve the appearance of the piece: The plating metal (nickel) is connected to the anode (positively charged electrode) of the electrical circuit, while the AM part is placed at the cathode (negatively charged electrode). Both are immersed in a specially developed electrolytic solution (bath). At this point, a DC current is supplied to the anode, which oxidizes the metal atoms in the nickel and

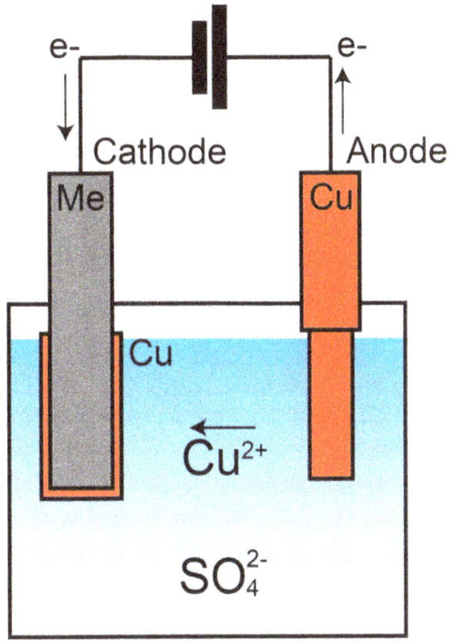

Fig. 12.22 The electroplating process

dissolves them into the bath. The dissolved nickel ions are attracted to the cathode and deposited (plated) onto the AM part (Fig. 12.22).

Most commonly, however, plating is done with several layers of different metals. The first layer of plating is often done with copper, as it is relatively easy to polish to a highly smooth surface. The copper layer is then plated with a subsequent layer such as nickel, chrome, silver, etc.

Factors that impact the final plating result include:

- the chemical composition and temperature of the bath;
- the voltage level of the electric current;
- the distance between the anode and the cathode;
- the electrical current application's length of time.

As with anodizing, with AM parts, you have more freedom to design in better hanging points for when the part is hung in the electroplating bath.

12.3.8 *Painting*

For painting metal AM parts, refer to the section on painting AM polymer parts, as the same principles apply. The only difference is that you may need to use a sanding primer that is suitable for metal.

Additionally, powder coating is possible with metal parts. Powder coating works by electrostatically depositing a dry powder on the surface of the part. The powder is then cured in an oven to allow the powder to melt and form a polymer skin on the product. Powder coating typically results in more durable coatings than wet painting.

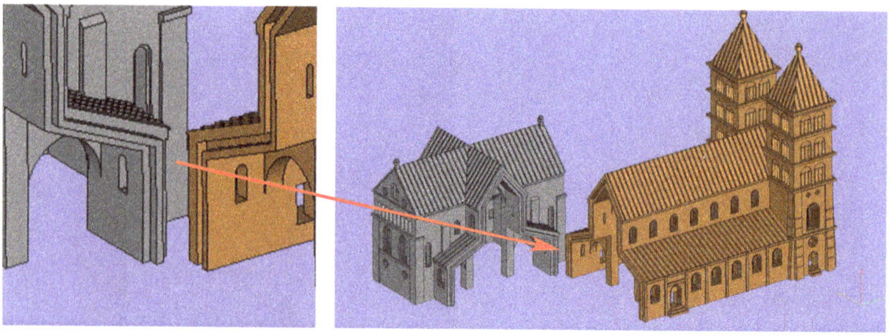

Fig. 12.23 Always include some kind of male/female joint between parts that need to be glued together

12.4 Gluing and Welding AM Parts

It is common to need parts that are larger than can be produced on many AM systems. In these cases, the parts must be printed in several smaller pieces and then joined together. The most common way of doing this is by gluing them together. This applies equally to polymer and to metal parts.

With the majority of AM processes and materials, most commonly available epoxy glues will work. With some AM materials, cyanoacrylate glues (superglue) can also be used. With some metal, there are speciality adhesives on the market that may provide better glue joint properties. And, for metal parts, the separate parts can also be welded together.

For parts in ABS, applying a thin coat of acetone to the faces to be glued, and then pushing them together can form a very strong welded plastic bond. The acetone will dissolve the surface layer and allow the pieces to be joined without glue.

If parts do need to be glued or welded together, it is strongly recommended that male/female joints be added to the parts so that things line up nicely. This simple design change can greatly accelerate, and improve the quality, of the parts when they are glued together (Fig. 12.23).

12.5 Heat Treatment and Aging

12.5.1 Residual Stress Relief

In metal AM, residual stress is unavoidable, and effort must be made to design AM parts to minimize this. Operational strategies, such as scan patterns, which are discussed in the chapter on designing for metal AM, can also be used during printing to minimize residual stress.

As AM is a form of micro-welding, residual stress is the result of the rapid heating and cooling that is inherent to laser powder bed fusion. Each new layer of the part is created by moving the laser beam across the bed, and melting the top layer of powder and, at the same time, fusing it to the layer below. Heat flows from the hot melt pool into the solid metal below, and helps the molten metal to cool and solidify. As the laser beam is extremely small, this happens in a matter of micro-seconds. As a new layer of metal cools down solidifies on top of the previous layer, it contracts. The new metal is constrained by the solid structure below it and so, when it contracts, it sets up shear forces between the layers. In the case of AM, this residual tress can be so severe that the entire build plate can be bent by the forces of the part acting on it.

Stress relieving must be carried out on all AM metal products in order to minimise residual stresses in the structure, and thereby reducing the risk of dimensional changes during further manufacturing or final use of the component. With AM, this stress relieving is done as the first post-processing step after the build platform is removed from the machine. It happens while the parts are still attached to the build plate by their support material.

Stress relieving does not change the material's structure and does not significantly affect its hardness.

The stress relieving temperature, for steel parts, for example, is normally between 550 and 650 °C, and the parts are heated slowly over about one to two hours. Soaking time is then several hours, and varies depending on the size/mass of the part. A general rule-of-thumb is to soak for 1 h per 25 mm of material thickness. After the soaking time the components should be cooled down slowly in the furnace, to about 300 °C, and then can finish cooling in air. A slow cooling speed is important to avoid tensions being reintroduced into the part by temperature differences in different areas of the material. This is especially important when stress relieving larger components. For stainless steels, a higher temperature solution heat treatment is normally necessary than for steel.

Temperatures and times vary greatly from metal to metal and part to part. But the general principle is relatively straight forward: The idea is to evenly heat the part, and let it soak until the entire part, both thick sections and thin sections, have reached a temperature equilibrium. After that, the idea is to slowly cool down the part, so that the thick sections and thin sections cool down at exactly the same rate. If the thin sections were to cool down quicker than the thick sections, for example, then residual stress would be reintroduced into the part.

If necessary, stress relieving can be performed in a furnace with protective gas to protect surfaces from oxidation. In extreme conditions, vacuum furnaces can be used.

12.5.2 Hot Isostatic Pressing (HIP)

Hot isostatic pressing (HIP) is a form of heat treatment that uses high pressure to improve material properties. That pressure is applied by an inert gas, usually argon.

Time at elevated temperature and pressure allows plastic deformation, creep and diffusion to occur. AM parts for critical applications are HIPed to eliminate internal microporosities, thereby improving mechanical properties by removing defects. It also improves the material's mechanical properties and workability. Metal parts that have been HIPed can reach metallurgical properties similar to wrought or forged parts.

The HIP process subjects a component to elevated temperature and gas pressure in a high pressure containment vessel. The pressurizing gas most widely used is argon (inert gas is used, so that the material does not react chemically). The chamber is heated, causing the pressure inside the vessel to increase. Pressure is applied to the material from all directions (hence the term "isostatic").

Simply put, HIP involves squeezing the part equally from all directions under high temperature in order to improve its properties. The pressure and temperature will eliminate most of the porosities or surface micro-cracks. Because the part is being squeezed equally from all directions, including any internal faces that are accessible to the gas, the process has relatively little effect on the dimensions of the part (though this must, of course, be verified for each particular application) (Fig. 12.24).

12.5.3 Case Hardening and Gas Nitride Treatment

Gas nitriding is a thermochemical case hardening process used to increase wear resistance, surface hardness and fatigue life by dissolution of nitrogen and hard nitride precipitations. It is a low temperature (typically 520 °C/970 °F), low distortion "thermochemical" heat treatment surface hardening process, where nitrogen is added to the surface of metal parts using dissociated ammonia as the source. Gas nitriding develops a very hard case in a component at relatively low temperature, without the need for quenching.

Favoured for components that are subjected to heavy loading, nitriding imparts a high surface hardness which promotes high resistance to wear, scuffing, galling and seizure. Fatigue strength is increased mainly by the development of surface compressive stresses. The wide range of possible temperatures and case depths, which allow adjustment of different properties of the treated parts, give gas nitriding a broad field of applications.

Nitriding is most effective when applied to the range of steels containing nitride-forming elements such as chromium, molybdenum, vanadium and aluminium. The process is also applicable to tool steels such as hot-work, cold-work and mould steels. In general, all ferrous materials can be gas nitrided up to 5% chromium. For higher contents of alloying elements and for gas nitriding of stainless steel, plasma nitriding might be considered.

For optimum results, the material should already be in a hardened and tempered condition prior to gas nitriding. Note also that gas nitriding of sintered steels with low density is not recommended.

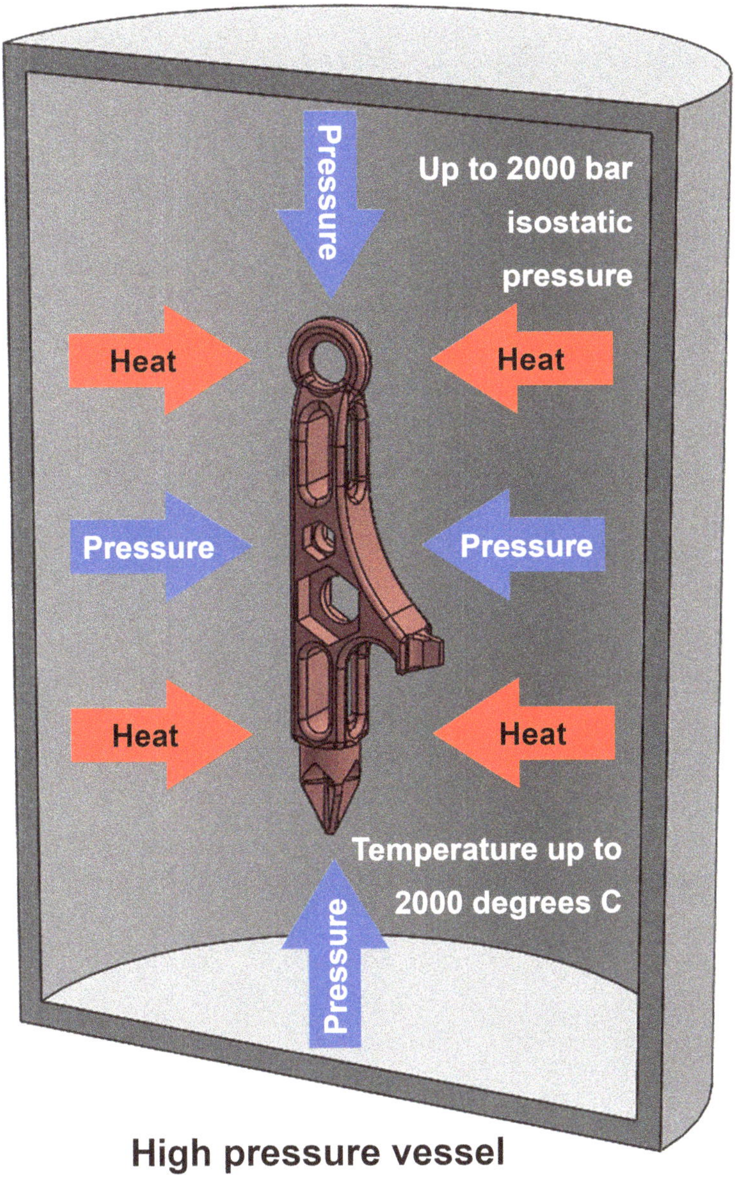

Fig. 12.24 The HIP process

Chapter 13
The Future of Additive Manufacturing

Additive manufacturing is developing very rapidly. Every few months we see new technologies, new materials, new software, and new AM products coming to market. It is of great importance to those with an interest in AM to keep abreast of some of the upcoming developments as they will, without doubt, affect how we develop future products.

If one examines where the bulk of research and development effort is being put today, it is, by far, focused on speed: Developing machines that are faster to eventually meet the need for the increased changes in volumes required by industry. There is also much research in new materials that are specifically designed for AM. This is both to have materials with better mechanical properties and to have materials that can be processed faster by the machines.

Another area of growth is in larger machines. The largest metal machines today, for example, have build volumes of around 500 mm × 500 mm × 500 mm. But developments are underway, using a number of the different AM technologies discussed in this book, to develop systems with much larger build volumes. The Aeroswift machine, for example, by South African company Aerosud is expected to have a build volume of 2 m × 0.6 m × 0.6 m.

Some other area of upcoming development that are of interest are discussed below.

13.1 Functionally Graded Materials

A particularly interesting area of materials growth is in functionally graded materials, smart material and multi-materials. Functionally graded materials are materials that change properties throughout a part. The functionality gradient can occur either because the actual material changes across the spectrum of the part, or because the geometry changes within the part (Fig. 13.1).

Smart materials are materials that can change properties in response to an external stimulus, and multi-materials are related to printers that can print in several different

O. Diegel et al., *A Practical Guide to Design for Additive Manufacturing*, Springer Series in Advanced Manufacturing,
https://doi.org/10.1007/978-981-13-8281-9_13

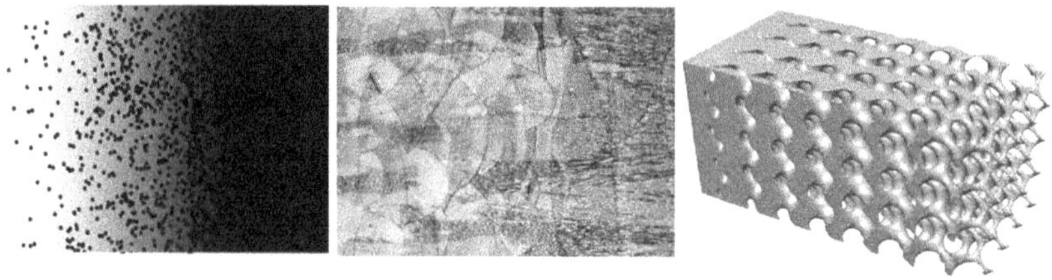

Fig. 13.1 Classification of different types of functionally graded materials, ranging from transitions between different materials to structural geometry changes within the material, courtesy of Mahmoud and Elbestawi [4]

materials at the same time to better meet the needs of products that are made up of multiple materials.

The new materials present tremendous potential to develop new smart products with improved functionalities. These materials are so new, however, that we don't quite know what to do with them yet, so novel applications will need to be developed at the same time that the materials are being developed.

13.2 Bioprinting

Tissue engineering and, ultimately, the printing of entire organs or body parts, is an active area of research worldwide. Researchers are currently able to bioprint a number of animal organs, human tracheas, heart valves, and bladders for humans that are engineered out of the patients own stem cells.

There are, currently two main techniques used for bioprinting. The first 3D prints a biodegradable polymer scaffold and seeds it with the patients stem cells. Over the incubation period, as the cells grow into what they are supposed to grow into, the polymer is degrading at the same rate that the cells are growing so that, by the time the cells have grown, all the polymer has been replaced by living cells. The other techniques is similar, but uses a hydrogel to suspend the stem cells in, and then print the mixture of hydrogel and stem cells (usually with extrusion or inkjet systems) into the right configuration. They then incubate them and, again, the cells gradually replace the hydrogel as they grow into what they are supposed to (Fig. 13.2).

The whole area of tissue engineering and bioprinting are still in their relative infancy, and it may be a few years before we see the ability to print more complex organs such as livers and hearts. But the potential it has for human health and well-being is tremendous, which makes it a truly valuable area of research.

It is important to understand that additive manufacturing is only a very small part of tissue engineering and bioprinting research. It relies as much on other disciplines including medicine, biology, software, etc. as it does on AM. It is also an area that will need the parallel development of a whole new branch of research into ethical and the social implications of such technologies.

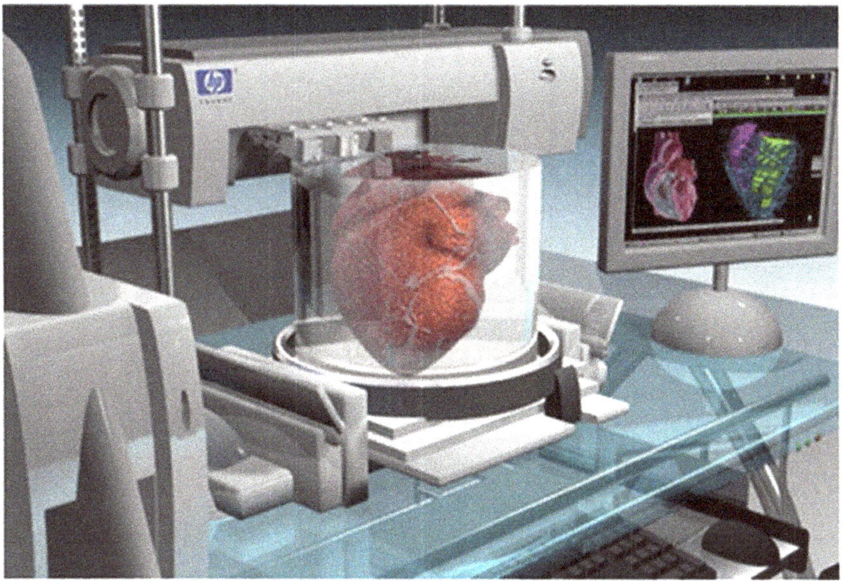

Fig. 13.2 Bioprinting may, one day, give the ability to print entire replacement organs, courtesy of Centromere121, CC BY-SA 3.0

13.3 Construction Applications

Work is going on at many universities and companies around the world to develop technologies capable of printing entire buildings and houses. The vast majority of the efforts in this area use a material extrusion based process in which concrete is extruded out of a nozzle and deposited. The concrete material can contain a variety of fillers, polystyrene, or fibers to reinforce it (Fig. 13.3).

Other notable efforts in this area include those of D-Shape, who are printing in reconstituted stone, using a binder jetting process, in which a powdered material is deposited as a layer, and a 'binder' is then printed onto the powder to create a reconstituted stone-like material. After the print is complete, all the loose remaining powder is vacuumed away, and the printed structure is left standing.

It has been predicted that, once the various house printing technologies have been fully developed, it will be possible to print an entire house within the timespan of only a few days. This includes not only the concrete structure, but all the amenities that form part of a functional house. It will be possible to order a house online, modify it to suits one's own design preferences, have the house printing machine trucked and installed on the construction site, and begin printing the house. A few days later and the house will be ready to move into.

As a reality check, however, one must remember that the already well established pre-cast method of building houses, in which sections of the house are pre-cast in a factory, and then assembled on site can do the same thing if one ignores the supply

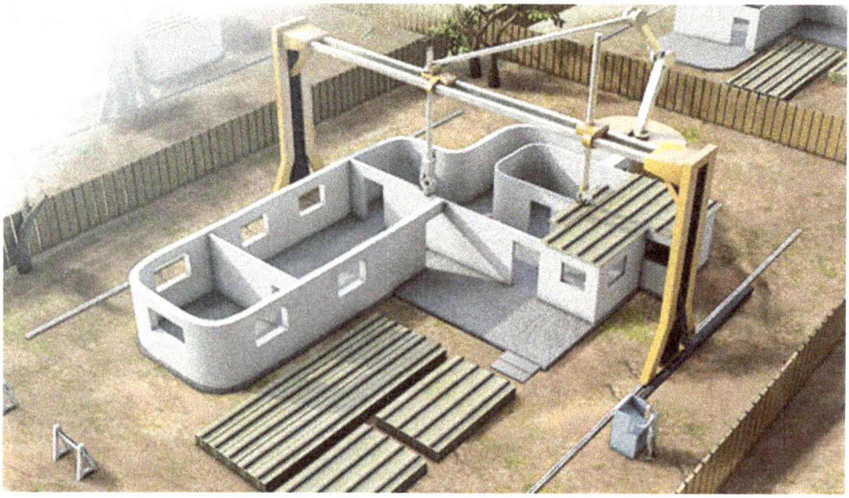

Fig. 13.3 Printing houses: the University of Southern California's Contour Crafting system, courtesy of Contour Crafting Corporation

chain aspects that add time to the build. But these would apply equally to 3D printed houses.

To date, the vast majority of 'printed houses' have been relatively simple structures that wold be possible to pre-cast more cost-effectively, which means that additive manufacturing has added little value to the end product. So, if 3D printed houses are to take off beyond just being a cool application of the technology, greater input from architects may be needed so as to design architectural structures that would not be possible to pre-cast. And applications need to be found in which AM allows buildings to be printed in a way, or in a context, that would otherwise not be possible. Otherwise additive manufacturing will add relatively little value to the construction business.

There are, however, some future areas of applications for the technologies that could be of great interest. One, in particular, is in printing space habitats. As exploration of the solar system continues, 3D printing could be a viable technology to mine raw materials from the planet being explored, and use that raw material to print habitats on the planet. This would alleviate the inefficient and very expensive challenge of sending up construction materials into space by rocket (Fig. 13.4).

13.4 Printed Electronics

From a product development point of view, conductive AM and printed electronics have the potential to truly change how we design products. Today, almost all electro-mechanical products are designed around rectangular flat circuit boards, so most products are largely rectangular in shape. Imagine no longer having this constraint,

Fig. 13.4 Printing habitats in space, courtesy of Contour Crafting Corporation

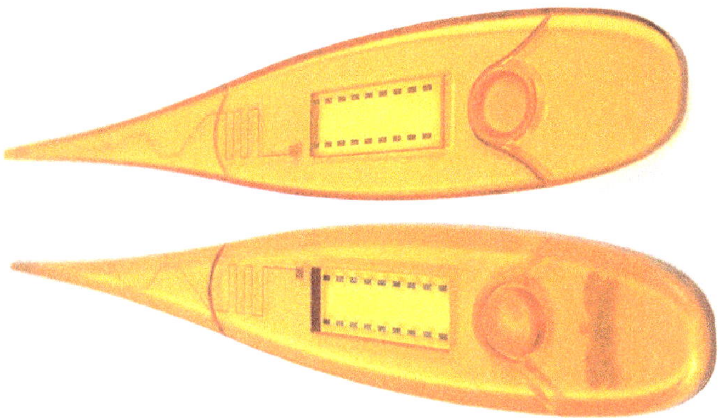

Fig. 13.5 Printed thermometer with printed electronic circuits, courtesy of Nano Dimension

with more freedom of design afforded by curved electronics and eliminating the need for wires (Fig. 13.5).

The level of readiness for the technology, today, is largely based around the ability to print wires integral to a polymer component, and printing circuit boards. Several companies have done substantial work in this area, including Voxel8, Nano Dimension, Optomec, and HP. This ability to print wires and circuit boards, alone, can already be of great benefit to product developers.

Some researchers have also proposed the idea of curved layer material extrusion as a way of eliminating the potential risk of discontinuous circuits in flat-layer applications. Much like the mechanical anisotropy caused by the joint between layers,

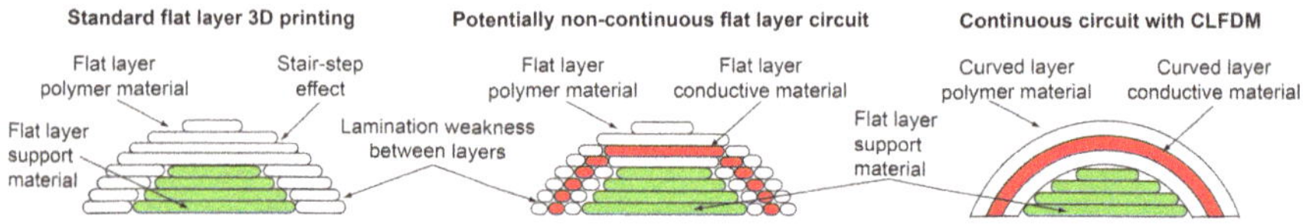

Fig. 13.6 Curved layer material extrusion for 3D printed electronics

with conductive 3D printing, the risk is that the conductivity between the layers may not be as good as the conductivity within the layer. The idea with curved layer FDM (CLFDM) is to, first, print all the support material, and then print the polymer material and conductive polymer material as curved layers on top of the support material. This ensures that the conductive tracks that form the wires are continuous and have the same conductivity throughout the wire (Fig. 13.6).

Researchers have also successfully printed relatively simple components like transistors and capacitors. But, once we reach the level of being able to print more advanced electronics, such as integrated circuit, the product development industry will truly be revolutionized.

13.5 Nano Printing

Though we are not quite yet using additive manufacturing at a nano-scale, researchers have been coming very close to this.

The pioneering work, in 2001, in femtosecond laser manufacturing of Prof. Satoshi Kawata, at Osaka University in Japan, for example has printed parts that are at the few microns scale. This includes parts, such as the nano-bull, which measure roughly one tenth of a human hair in size. Researchers at the Vienna Institute of Technology have also created a tiny Formula One race car, using two-photon lithography, which uses highly-focused light beams to manipulate then harden the resin molecules in exactly the right position. The 'two-photon' part of the name refers to how the resin only hardens when two photons hit it at the same time. The image is a great example of small scale printing and shows a printed cathedral at the very top of a pencil (Fig. 13.7).

This technology could, ultimately, mean the manufacture of nanobots that could, for example, be injected into the blood stream with the task of cleaning the blood stream of all undesirable molecules.

13.6 Food Printers

There is now a whole new field of research concentrating on developing 3D printers capable of printing food (Fig. 13.8). By far, the majority of the efforts in this area use material extrusion technologies to extrude food pastes in order to reconstitute

Fig. 13.7 Nanoscale 3D printed structure of the Sutyagin house, or "Spitzen-Forschung" (holiday house) in Arkhangelsk, Russia, on the tip of a pencil, made with two-photon polymerization, courtesy of MaeID—FutureRetrospectiveNarrative (Daniela Mitterberger, Tiziano Derme) and Technische Universitaet Wien (Aleksandr Ovsianikov and Peter Gruber)

facsimiles of food. So these machines are not so much 'printing food' as 'printing with food'. Some others use binder jetting technology to bind powdered materials, such as sugar, together with an ink-jetted binder.

One of the few efforts in true printing of food has been in printing meat, using bovine stem cells and a method similar to that described in the section on bioprinting above.

Much of the 3D printed food, so far, has focused on printing geometrically complex, or customized, foods such as chocolates, pasta, sugar wedding cake toppers and candies, etc. Though this is, of course, interesting, one can question whether it adds enough value to the food to be worth the relatively slow print speed and, therefore, whether it is a cost-effective way of producing food.

An area where 3D printed food can, however, potentially add great value is on printing customized foods for people with dysphagia, or other chewing and swallowing disorders. There is potential, for example, for printing food with internal lattice structures that would rapidly dissolve when put in the mouth, or be easy to chew and swallow by the customer. So could one print a carrot that looks like a carrot, tastes like a carrot, but is easy for the customer to chew or swallow? Not only could it have the right texture and consistency for the customer, but it could also have its ingredients customized to meet the users' needs with, for example, extra vitamin B or calcium (Fig. 13.9).

Fig. 13.8 The Cornucopia project food printer, courtesy of Zoran [7]

Fig. 13.9 Food for the elderly, courtesy of BioZoon

Chapter 14
Concluding Remarks

Additive manufacturing is in an extremely rapid state of flux. By the time you have read this book, new AM technologies, materials or software will have appeared on the market. It is a growing area of research and there are few universities that do not have some research program in AM, and existing and new companies are increasingly adding it to their production systems.

Some of the things that are needed to further increase its adoption include **more materials, faster AM systems, better surface finishes, and certifiable processes**. We also need better software tools to ensure that every part is printed correctly the first time based on customer requirements, and better design tools that will allow us to design safe products that are optimized for AM. And, in particular, we need to update our engineering and design education programs to include design for AM.

One fact that will not change in the near future will be the need to design products that wish to employ AM as part of their production process to take advantage of the benefits that AM can give. For this reason, it is essential that our upcoming engineers and designers be educated in design for additive manufacturing.

Some of the core aspects of such an education would, at the very least, include some of the following:

- Understanding of why and when to use AM.
- Understanding of how AM materials are made, and what the final material and mechanical properties of AM parts are.
- Hands-on operation of AM technologies to really understand the processes.
- A real understanding of AM post-processing.
- DfAM guidelines (wall-thicknesses, hole sizes, options for solid or sparse builds, tolerances, etc.).
- AM CAD tools, e.g: topology optimization and lattice structures, finite element analysis tools for AM, etc.
- And an understanding of the AM certification process.

O. Diegel et al., *A Practical Guide to Design for Additive Manufacturing*, Springer Series in Advanced Manufacturing,
https://doi.org/10.1007/978-981-13-8281-9_14

Some other questions that have not been addressed in this book, and will, no doubt, become future areas of research include:

- What happens to our business models and structures with the implementation of AM?
- Will the value chain for future products shift from manufacturing to design?
- How is our health and well-being affected by AM?
- What ethical questions does AM raise?
- What happens to our leisure time and lifestyles?

As this book is a compilation of additive manufacturing knowledge from a great many sources, it is very much a work in progress. Any input from readers is most welcome as it will help us improve future editions of this book.

Glossary of Terms

The following are key terms and abbreviations used in the field of additive manufacturing and 3D printing. Most of the terms in this glossary of terms (denoted with a *) and book conform to the ISO/ASTM 52900 terminology standard.

3D printer* Machine used for 3D printing.

3D printing* Fabrication of objects through the deposition of a material using a print head, nozzle, or another printer technology. Term often used in a non-technical context synonymously with additive manufacturing; until present times, this term has in particular been associated with machines that are low end in price and/or overall capability.

3D digitizing Same as 3D scanning.

3D scanning Metrological method of determining the size and shape of an object using some degree of automation; often involves an optical device, such as a laser, and sensors that calculate xyz coordinates using a technique called triangulation.

Additive layer manufacturing Same as additive manufacturing.

Additive fabrication Same as additive manufacturing.

Additive manufacturing* Process of joining materials to make parts from 3D model data, usually layer upon layer, as opposed to subtractive manufacturing and formative methodologies; historical terms are additive fabrication, additive processes, additive techniques, additive layer manufacturing, layer manufacturing, solid freeform fabrication, and freeform fabrication.

O. Diegel et al., *A Practical Guide to Design for Additive Manufacturing*, Springer Series in Advanced Manufacturing,
https://doi.org/10.1007/978-981-13-8281-9

Additive process	Same as additive manufacturing.
Additive system*	Additive manufacturing system, additive manufacturing equipment, machine and auxiliary equipment used for additive manufacturing.
Bed	Another name for an AM system build plate. This is the flat surface on which the parts are made.
Binder jetting*	Additive manufacturing process in which a liquid bonding agent is selectively deposited to join powder materials.
Build plate	The area on which an additively manufactured part is printed.
Build platform	Same as build plate.
Build chamber*	Enclosed location within the additive manufacturing system where the parts are fabricated.
Build envelope*	Largest external dimensions of the x-, y-, and z-axes within the build space where parts can be fabricated.
Build volume*	Total usable volume available in the machine for building parts.
Digital light processing	A display device that creates an image using an array of micromirrors; each mirror represents one or more pixels in the projected image.
Direct digital manufacturing	Direct digital manufacturing: The production of final products or components using AM technology.
Directed energy deposition*	Additive manufacturing process in which focused thermal energy is used to fuse materials by melting as they are being deposited. "Focused thermal energy" means that an energy source (e.g., laser, electron beam, or plasma arc) is focused to melt the materials being deposited.
Direct metal deposition	A trade name used by DM3D for the company's directed energy deposition technology.
Direct metal laser sintering	A trade name used by EOS for the company's metal powder bed fusion technology.
Electron beam melting	A trade name used by Arcam for the company's electron beam-based metal powder bed fusion technology.
Extruder	The 3D printer component which melts and deposits the melted plastic.

Facet*	Typically a three- or four-sided polygon that represents an element of a 3D polygonal mesh surface or model. Triangular facets are used in the file formats most significant to AM: AMF and STL files; however, AMF files permits a triangular facet to be curved.
Filament	The polymer filament used in most of the material extrusion processes, which melts and is deposited by the extruder.
Fused deposition modeling	A trade name used by Stratasys for the company's material extrusion technology.
Fusion*	Act of uniting two or more units of material into a single unit of material.
Hot isostatic pressing	Uses heat and isostatic pressure to reduce or eliminate the porosity in metals and increase the density of ceramics.
Hybrid manufacturing system	Manufacturing system that uses both additive and subtractive technologies.
Laser sintering	Powder bed fusion, particularly as it applies to polymers.
Layer additive manufacturing	Same as additive manufacturing.
Maker	A member of a technology-based do-it-yourself (DIY) community.
Material extrusion*	Additive manufacturing process in which material is selectively dispensed through a nozzle or orifice.
Material jetting*	Additive manufacturing process in which droplets of build material are selectively deposited. Example materials include photopolymer and wax.
Metrology	Science of measurement.
Near net shape*	Condition where the components require little post-processing to meet dimensional tolerance.
Photopolymer	A thermoset polymer that changes properties when exposed to ultraviolet or visible light; typically, a photopolymer changes from liquid to solid during photopolymerization.
Powder bed fusion*	Additive manufacturing process in which thermal energy selectively fuses regions of a powder bed.
Post-processing*	One or more process steps taken after the completion of an additive manufacturing build cycle in order to achieve the desired properties in the final product.

Process parameters* Set of operating parameters and system settings used during a single build cycle.

Prototype* Physical representation of all or a component of a product that, although limited in some way, can be used for analysis, design, and evaluation.

Prototype tooling* Molds, dies, and other devices used for prototyping purposes; sometimes referred to as bridge tooling or soft tooling.

Reverse engineering A method of creating a digital representation from a physical object to define its shape, dimensions, and internal and external features.

Rapid prototyping* Application of additive manufacturing intended for reducing the time needed for producing prototypes. Historically, rapid prototyping (RP) was the first commercially significant application for additive manufacturing, and has therefore been commonly used as a general term for this type of technology.

Rapid tooling* Application of additive manufacturing intended for the production of tools or tooling components with reduced lead times as compared to conventional tooling. Rapid tooling may be produced directly by the additive manufacturing process or indirectly by producing patterns that are in turn used in a secondary process to produce the actual tools.

Selective laser melting A generic name for metal powder bed fusion.

Selective laser sintering A trade name used by 3D Systems for the company's polymer powder bed fusion technology.

Sheet lamination* Additive manufacturing process in which sheets of material are bonded to form a part.

SMEs Small and medium-sized enterprises.

Solid model 3D CAD representation somewhat analogous to using material, such as wood or plastic, to create a shape. Many solid-modeling software products use geometric primitives, such as cylinders and spheres, and features such as holes and slots, to construct 3D shapes. Solid models are preferred over surface models for additive manufacturing because they define a closed, "water tight" volume—a requirement of most additive manufacturing systems.

Support material	Additional material which is used to support overhanging features that could not otherwise be printed, or to act as a heatsink to remove heat from the part being printed. Support material is, in most cases, automatically generated in the correspondent AM machine software.
Surface model*	Mathematical or digital representation of an object as a set of planar or curved surfaces, or both, that can, but not necessarily have to represent a closed volume.
Topology optimisation	The use of mathematics to optimise the strength-to-weight ratio of a design. The approach minimizes the use for a given set of load and constraint conditions.
Topological optimisation	Same as topology optimisation.
Tool, tooling	Mold, die, or another device used in various manufacturing processes such as plastic injection molding, thermoforming, blow molding, vacuum casting, die casting, sheet metal stamping, hydroforming, forging, composite layup, machining, and assembly fixtures.
Triangulation	Method of inferring the location of a point on a surface by projecting light onto the surface and observing that light from a different angle or orientation.
Ultem	Common material extrusion technology material. Ultem, a polyetherimide is an amorphous, amber-to-transparent thermoplastic with characteristics similar to the related plastic PEEK. Ultem resins are often used in medical and chemical instrumentation due to their heat resistance, solvent resistance and flame resistance.
Vat photopolymerization*	Additive manufacturing process in which liquid photopolymer in a vat is selectively cured by light-activated polymerization.
Voxel	Volume element; objects and three-dimensional datasets can be divided into an array of discrete elements, called voxels, on a regular grid in three-dimensional space.

* denotes ISO/ASTM 52900 standard definition.

References and Further Reading

This section contains a list of references used in this book as well as a number of good sources for further reading that will provide a wealth of information on additive manufacturing.

Further Reading

Chua CK, Leong KF (2014) 3D printing and additive manufacturing: principles and applications. World Scientific

Redwood B, Garret B, Schöffer F, Fadell T (2018) 3D printing handbook. https://www.3dhubs.com/3d-printing-handbook

Gibson I, Rosen D, Stucker B (2015) Additive manufacturing technologies: 3D printing, rapid prototyping, and direct digital manufacturing, 2nd edn. Springer, Berlin

Lipson H, Kurman M (2012) Fabricated: the new world of 3D printing. Wiley, New York. ISBN: 978-1118350638

Beaman J et al (1997) Solid freeform fabrication: a new direction in manufacturing. Springer, Berlin

Wohlers Report 2019: 3D printing and additive manufacturing state of the industry. www.wohlersassociates.com (2019). ISBN: 978-0-9913332-5-7

www.3dhubs.com/knowledge-base

www.shapeways.com/tutorials

https://i.materialise.com/en/tutorials

www.sculpteo.com/en/materials/materials-design-guidelines/

European Powder Metallurgy Association: www.epma.com/european-additive-manufacturing-group

www.renishaw.com/en/am-guide–41140

O. Diegel et al., *A Practical Guide to Design for Additive Manufacturing*, Springer Series in Advanced Manufacturing,
https://doi.org/10.1007/978-981-13-8281-9

References

1. Manriquez Frayre A, Bourell DL (1990) Selective laser sintering of binary metallic powder. In: Proceedings of the solid freeform fabrication symposium. The University of Texas Mechanical Engineering Department, 6–8 Aug 1990, pp. 99–106.
2. Bourell DL, Beaman JJ Jr, Leu MC, Rosen DW (2009) A brief history of additive manufacturing and the 2009 roadmap for additive manufacturing: looking back and looking ahead. In: Fidan I, Calisir F (eds) Proceedings of the US-Turkey workshop on rapid technologies, 24–25 Sept 2009. Tübítak, The Scientific and Technical Research Council of Turkey, Istanbul, 2009, pp. 5–11.
3. Haberland C (2012) Additive Verarbeitung von NiTi-Formgedächtniswerkstoffen mittels Selective Laser Melting (Additive manufacturing of NiTi shape memory alloys by means of selective laser melting), Herzogenrath: Shaker Verlag GmbH. ISBN: 978-3-8440-1522-5.
4. Mahmoud D, Elbestawi MA (2017) Lattice structures and functionally graded materials applications in additive manufacturing of orthopedic implants: a review. J Manuf Mater Process 1(2):13.
5. Wright S, Arcam EBM Guide, CC BY 4.0, https://github.com/Gongkai-AM/Machine-Guides/blob/master/Arcam%20EBM%20Guide.md
6. Zoran A (2011) The 3D printed flute: digital fabrication and design of musical instruments. J New Music Res (JNMR) 40(4):379–387.
7. Zoran A, Coelho M (2011) Cornucopia: the concept of digital gastronomy. Leonardo J Arts Sci Technol 44(5):425–431.

GPSR Compliance
The European Union's (EU) General Product Safety Regulation (GPSR) is a set of rules that requires consumer products to be safe and our obligations to ensure this.

If you have any concerns about our products, you can contact us on

ProductSafety@springernature.com

In case Publisher is established outside the EU, the EU authorized representative is:

Springer Nature Customer Service Center GmbH
Europaplatz 3
69115 Heidelberg, Germany

www.ingramcontent.com/pod-product-compliance
Lightning Source LLC
Chambersburg PA
CBHW061445110126
38050CB00003B/55
9789811382802